Thomas Ott · Frank Swiaczny

Time-Integrative Geographic Information Systems

Springer
Berlin
Heidelberg
New York
Barcelona
Hong Kong
London
Milan
Paris
Singapore
Tokyo

Thomas Ott · Frank Swiaczny

Time-Integrative Geographic Information Systems

Management and Analysis of Spatio-Temporal Data

With 99 Figures

Springer

Dr. Thomas Ott
Dipl. Geogr. Frank Swiaczny
University of Mannheim
Department of Geography
L9, 1–2
68131 Mannheim
Germany

ISBN 3-540-41016-3 Springer-Verlag Berlin Heidelberg New York

Cataloging-in-Publication Data applied for
Die Deutsche Bibliothek – CIP-Einheitsaufnahme
Ott, Thomas: Time-Integrative Geographic Information Systems: Management and Analysis of Spatio-Temporal Data / Thomas Ott, Frank Swiaczny. – Berlin; Heidelberg; New York; Barcelona; Hong Kong; London; Milan; Paris; Singapore; Tokyo: Springer, 2001
ISBN 3-540-41016-3

Springer-Verlag Berlin Heidelberg New York
a member of BertelsmannSpringer Science+Business Media GmbH

Printed in Germany

Cover-Design: Erich Kirchner, Heidelberg
Typesetting: Camera-ready by the authors

SPIN 10726470 30/3130/xz – 5 4 3 2 1 0 – Printed on acid-free paper

Preface

This book is the provisional result of more than 10 years of continued discussion with friends and colleagues from neighbouring disciplines. Although only a small minority of the millions of GIS users on this planet are geographers, it seems that somehow, geographers are a kind of natural contact persons for historians, archaeologists, economists, social scientists or others who are looking for appropriate ways of working with spatial data.

We received constant encouragements and many valuable suggestions from our colleagues. Particularly we wish to thank the members of the GIS Study Group of the German Association of Geography (AK GIS) as well as the participants of a workshop in June 2000 on "Mapping Europe's historic boundaries and borders" which was generously sponsored by the European Science Foundation. Among the individuals we owe special appreciation are Humphrey Southall and Ian Gregory (The Great Britain Historical GIS Programme, University of Portsmouth), Michael Goerke (European University Institute, Florence), Konrad Pierau (Center for Historical Social Research, University of Cologne), Bernhard Holfter (Förderverein Historische Grundkarte, Leipzig) and Stephan Riediger (Department of History, University of Mannheim).

A considerable portion of work and insight was provided by Frank Hellwig who open-handedly shared the ideas of his dissertation. Furthermore we owe gratitude to our superiors Christoph Jentsch and Paul Gans which accompanied our work with a perfect mixture of autonomy and attention. Our list would not be complete if we did not thank Mareike Ott for her always accurate and expeditious collaboration. Finally, we thank our wives Bärbel and Renate for putting up with our erratic but intensive work pattern during the project.

Several illustrations in this book are loosely based on the work of others; we rely on source citations within the figure captions as our acknowledgement. We also point out that some figures have appeared in earlier publications as well as in our printed classroom materials.

Throughout this book we refer to several commercial products. Our mention does neither constitute an endorsement nor a refutation; we rather refer to these brand names as illustrations and examples for different approaches or concepts. Even though some details mentioned may be outdated soon, we offer these examples in order to assist the transfer of the theoretical concepts to the practical world. We share the belief that the ongoing progress made by software developers all over the world is one of the exciting aspects of GIS technology.

This book is designed to provide general information on temporal Geographic Information Systems. We tried to make every effort to make this book complete, and as accurate as possible. While the assistance from all the sources helped create this book, none of them hold any responsibility for the shortcomings and errors it may still contain. Not being native speakers, lapses in the language are, however, another item. It was the decision of the publisher to make this book available to a wider readership. We seriously hope that neither you, the gentle reader, nor the people at Springer have to regret this decision.

To keep this book up-to-date we maintain a special website at http://www.time-gis.de. Please feel invited to visit the site and share your thoughts and comments on the book with us and others interested in time-integrative GIS.

Mannheim, August 2000

Thomas Ott
Frank Swiaczny

Contents

List of abbreviations

ACSM	American Congress of Surveying and Mapping
AGI	Association for geographic Information
DB	Data Base
DBMS	Data Base Management System
ER	Entity-Relationship
GIS	Geographic Information System
IGU	International Geographical Union
LCG	Least Common Geometry
NCGIA	National Center for Geographic Information and Analysis
OLE	Object Linking and Embedding
OO	Object Oriented
OODBMS	Object Oriented Data Base Management System
RDBMS	Relational Database Management System
SQL	Structured Query Languange
TSQL	Temporal SQL

List of figures

List of tables

1 Introduction

"And crawling on this planet's face, some insects called the human race. Lost in time. And lost in space."
(The Rocky Horror Picture Show 1975)

1.1 The relationship of space and time

Time has always been a fascinating dimension of human existence. It is also an aspect in the field of geographic information science that today is getting more attention than ever before. Although the concept of storing unique attribute information linked to objects within a GIS is fundamental, adding the time element raises some intriguing questions and presents new difficulties. Time adds process to pattern and the inclusion of temporal aspects within a GIS opens new perspectives. It is definitely one of the most exciting prospects that lies ahead in this field.

In general, a time-integrative or temporal GIS may be defined as an attempt to store and analyse spatial objects and changes in their attributes through time (Castagneri 1998). Thus we may distinguish three kinds of data elements to be processed in a temporal GIS: spatial, temporal and attribute data (Peuquet 1988; Openshaw 1994:88; see figure 1.1). The triangular arrangement (Peuquet 1988) is both, a means to reflect on geographic information in general and a method for structuring information in a GIS database. Taken together they represent the attributes of an object located at a certain position at a certain time.

Yet, two contrasting ways of thinking on the nature of space and time evolved since ancient Greece. While the *absolute perspective* looks at space and time as an expanse or container in which all things exist, the *relative perspective* perceives time and space as closely interrelated and interdependent. "More recently, the post-Einstienian [sic!] relativistic perspective now speaks of space-time, or the space-time manifold, in which space and time are mutually constituted and inseparable" (Harrower 1999: 4).

This book, however, follows a Newtonian perspective: time is understood as an independent dimension, nonetheless closely related and similar to the spatial dimensions. As a result, the objects and entities referred to in this book are thought of as being contained in the space-time continuum. Based on this grounding it would be possible to analyse space and time independently. However, the common metaphor of time as a line stretching indefinitely from past to present and

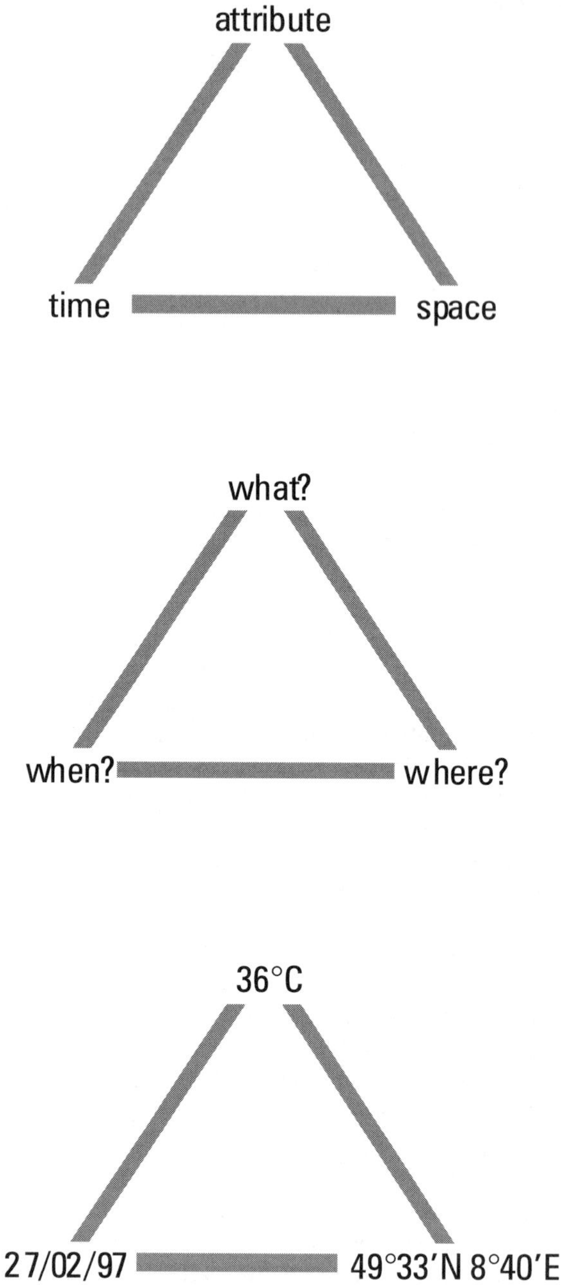

Fig. 1.1 Triadic model of space, time and attribute (adopted from Peuquet 1988)

into the future is an indication that the temporal and spatial dimensions have something in common. "The concept of simultaneity, for instance, is analogous to spatial identity. Furthermore, the data structures used to retrieve events that occurred within a certain month might be analogous to the ones needed to find the service station within the first 100 miles along a certain road" (Hermosilla 1994: 122).

The data which can be represented in a GIS may be classified in dimensions (Davis/Williams 1989). Commonly four dimensions – length, width, depth and time – are differentiated.[1] However, transferred to a GIS at least one additional dimension must be added for the attribute data to be analysed (see chapter 2.5). Hence, we may extend the triadic model of space-time-attribute into a multidimensional cube (see figure 1.2). the information processed in the GIS may exhibit zero, one, two, or three spatial dimensions and either zero or one temporal dimension (the data may have a temporal component or not). The third axis of the cube represents the measurement of attributes. As phenomena potentially possess many attributes, this axis could in theory be divided into *n* dimensions. However, we have to consider that attribute data may either be numerical (interval / ratio) or categorical (nominal / ordinal). Attribute data may therefore refer to an object itself, or to some measured characteristics of an object (Harrower 1999: 10-11).

Throughout the history of geography many workers have studied temporal progressions in an analytic manner. There are for example the studies of innovation and diffusion (Hägerstrand 1952), which in turn led to the ideas of "Cellular Geography" (Tobler 1979) and "cellular automata" (Couclelis 1985). Furthermore there is Hägerstrand's more qualitative study of the effects of time and space on people which he termed "Time Geography" (1970). Most of the emphasis in his Time Geography is upon the concept of society evolving over time, and this development being influenced by intersecting external activities.

Interestingly enough, in a way geographers are jointly responsible for the neglect of temporal aspects in GIS technology. Following the quantitative revolution of the 1960s geography was labelled as a "science of space". The temporal component was often reduced to the role of an attribute out of many. Maps depicting the location and distribution of spatial phenomena became both the means and the product of geographical analysis.[2]

Trying to remedy this undesired development, there has been a fair amount of research and debate on the subject of time-integrative GIS in the last 15 years, predominantly from a technical perspective (see the research bibliography in chapter 9.1). Many design proposals and even some intriguing prototype software

---

[1] This basic concept of modern physics was introduced by Hermann Minkowski. His idea of a four-dimensional space, combining the three dimensions of physical space with that of time, laid the mathematical foundation of Albert Einstein's general theory of relativity (see Minkowski & Hilbert 1911).

[2] "The primacy of spacer over time is due to the close association between cartography and geography. [...] Perhaps if cartography had allied with history, time would be treated with equal or greater import than space" (Harrower 1999: 5).

solutions were presented. However, not much has been accomplished in the real world, although potential applications of temporal GIS abound.

Most introductory texts on GIS (see section 1.3) still try to avoid the complex issues raised by the integration of time into GIS. There are, however, a few exceptions to the rule. Gail Langran (1992) wrote the seminal and still unparalleled work on "Time in Geographic Information Systems", reviewing and discussing many of the possibilities and technologies in detail. Other substantial contributions continue to come from the members of the GeoVISTA center at Penn State University with Donna Peuquet and Alan MacEachren in a prominent position. The discussion on temporal GIS has profited from the GISData initative sponsored by the European Science Foundation and the national research councils of 14 European countries.[3]

However, this book is not intended to be just another item on the aforementioned list. Thus, the targets and the potential readership of the book are described in the subsequent sections.

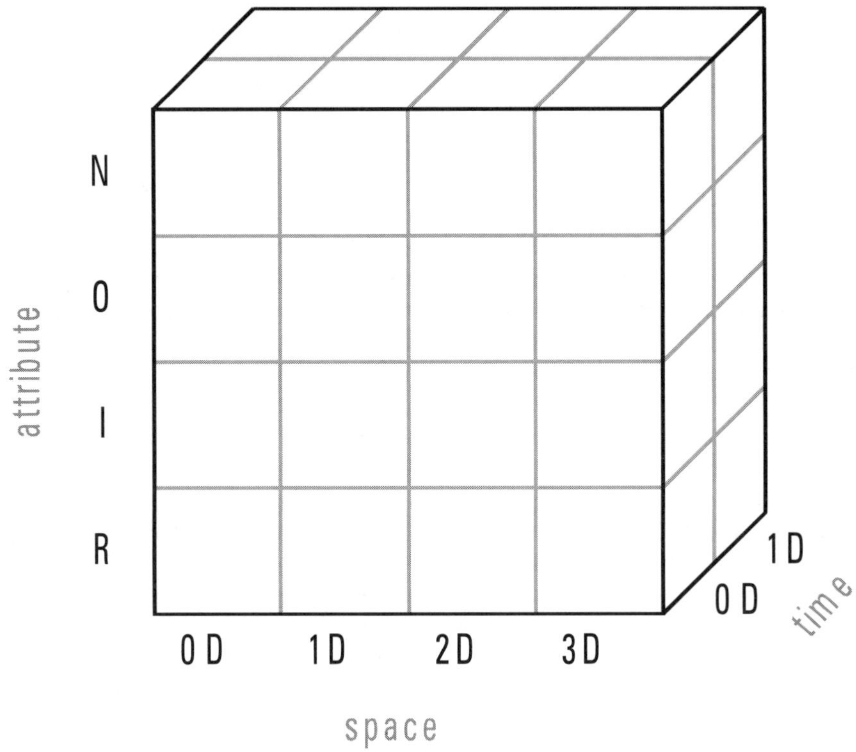

Fig. 1.2 Multidimensional cube of space, time and attribute (adopted from Harrower 1999:14)

---

[3] For a list of the people involved and the resulting series of publications please refer to the GISData website at http://www.shef.ac.uk/uni/academic/D-H/gis/gisdata.html

1.2 Targets and contents of the book

In the last years the use of Geographic Information Systems was no longer limited to geographers, spatial analysts, earth scientists or urban and regional planners. They have gained an increasing number of useful purposes in such dissimilar fields like marketing, logistics, archaeology, sociology or biology. As a result of the development of easy to use GIS and desktop-mapping software products the market share of these non-geographic applications shows a steady rise. Specialized software for the specific needs of these new GIS users has been developed based on existing GIS-packages. Furthermore the supply of spatial data – a precondition for the effective use of GIS technology – has expanded dramatically.

However, new GIS users are often surprised to discover that the software has great difficulties in answering questions such as, "When did the river flood more than 20% of the agricultural land?", "Were there ever any parcels of land with industrial zoning within the polluted area?", "Are there any significant connections between the extension of railway lines and the population development in the 19th century" or responding to requests such as, "Display the forest boundary changes between 1850 and 1950." An urban planner might be disappointed that the municipal GIS cannot simply report what the previous zoning was for a particular strip of land, historians or sociologists often fail by trying to compare statistical data or election results for shifting administrative units or voting districts (Candy 1995).

These setbacks are due to the fact that commercial GIS packages mostly lack the ability to perform temporal analysis of spatial data. Yet, not only historical research is dependent on the ability to represent and analyse temporal processes in a GIS. Today, the widespread application of GIS in urban, regional and environmental planning, in marketing and logistics, or scientific disciplines like archaeology requires the integration of historical data in order to perform time-change analyses or to deal with temporally variable data. Another crucial motivation for the development of time-integrative GIS techniques is the fact, that enormous amounts of data have been collected in the last decades by census offices, research projects and not the least companies, which can be revaluated and reused.

The development of appropriate methods and tools as well as their implementation in ready to use GIS packages is impeded by the fact that only very few countries or organizations have formulated a standard for geographic and/or attribute data with respect to temporal representation and analysis. Furthermore, on the part of the census offices very important groundwork in form of historic territorial databases is more or less missing. The same holds true for digital historic base maps covering all territories and epochs. Although there are numerous groups and individuals working on smaller stand alone projects, there are – with only a few exceptions to the rule – no higher institutions which work as a clearinghouse for coordination and cooperation.

1.2.1 What this book is about

Main target of this book is, to contribute to the ongoing discussion on temporal GIS and to present practical solutions for the conception and implementation of time-integrative geographic information systems. Main purpose of a temporal or time-integrative GIS is, to reproduce temporal processes or sequences of events of the real world in a model in such a way, to make them accessible for spatial query, analysis and visualization. For that it is necessary to link the spatial information with the corresponding spatial objects (geo-coding). In order to be able to represent and analyse both, the temporal and spatial interrelations, the modelling of spatio-temporal data (i.e. the process to derive useful information and "knowledge" from a set of raw data) requires extraordinary assiduousness.

On closer examination of the traditional formation of a layer GIS (see chapter 2.2) it is obvious that thematic, temporal or vertical information layers (entities) each form a separate category, which basically are not related to one another. Already with the construction of the data model it therefore has to be considered, which data layers are to be processed later in combination and which analysis methods (e.g. time sections vs. process analysis) are necessary. Furthermore, as subsequent modifications of an existing data structure are almost impossible, questions of scale or accuracy, temporal extension and smallest spatial units to be analysed (e.g. plots vs. administrative units) have to be taken into consideration right from the beginning.

An important prerequisite for the successful GIS use is therefore a systematic plan of action for the modelling of the significant questions and the gradual steps of abstraction from the conceptual model of the real world up to the physical data model in the computer. One must consider, which entities and which relations between these entities have to be represented in a model of the real world and which analysis approaches should be possible with this model. As an example, aggregate census data is quite frequently spatially referenced via the construct of the linkage between housing location and administrative unit. However, the selection of this linkage implies already certain assumptions on interactions and interrelations, which are themselves subject of examination.

The specific problems of a historical or time-integrative GIS are based on the necessity of simultaneous spatial and temporal reliability. If the individual information layers of the GIS are not to be explored independently, it is of fundamental importance, that on the level of the logical data model, identical points (even as element of a line or polygon) are represented by exactly one pair of coordinates. This holds true for identical points in different thematic layers as well as identical points in different temporal layers or different scales. In regard of the practice of data capturing the latter forms the greatest challenge. Historical information has to be digitised as time sections from diverse maps of differing precision and scale. In opposition to different thematic layers in a specific time section, which usually can be composed from base maps with coherent accuracy, the digitisation of historic sequences is impeded by differing map projections and scales as well as varying levels of precision. Although these problems are to be mastered, both in a fundamental and methodical way as well as in practice with an appropriate expenditure on programming, smaller projects are usually confronted with unsolvable difficulties.

The demand for spatial and temporal reliability leads to the most prominent problem with the construction of a historical GIS: the representation of temporal information in a conceptual data model and the reproduction of temporal data in a logical data model. The structured proceeding described in the book is not to be understood as an end in itself. It rather permits a first overview of the system to be analysed. The more exact the elements and interactions of the system are studied and transformed into a conceptual model against the background of theoretical knowledge, the easier and better it will be possible to form the logical entity-relationship model in the next step. Another advantage of this procedure results from the fact that missing links and relations can already be uncovered, before the actual data acquisition has started. Since the construction of the data objects exerts strong influence on the possibilities of analysis, their feasibility can already be checked at an early stage and modified if necessary. Furthermore the entity-relationship model forms a basis of decision for the selection of data sources, as it permits to identify inappropriate or redundant information and thus may save operating time and costs. However, the significance of these theoretical considerations is qualified by the characteristics of the available sources, since these are not available or accessible in every desired temporal and/or spatial resolution.

1.2.2 What this book is not about

From a theoretical point of view temporal GIS is a very hot topic on the research agenda. Many GIS data models have been proposed to incorporate temporal information into spatial databases, prototypes have been developed, and many issues have been discussed mainly in an information science perspective (see the research bibliography in chapter 9). This book offers no new insight or additions from this information science standpoint, it rather tries to translate the theoretical contributions others have made into recipes for everyday work. This intention may be disappointing to some readers, but we hope to provide a welcome aid for others, who are busy doing their project work and have no time or resources left to grapple with valuable but nevertheless avant-garde and complex theories.

Since this is an introductory book, some aspects are dealt with relatively superficially and others are only mentioned in passing. For example, there is little discussion of mathematical methods and formulas for handling the geometrical aspects. However, questions related to the introduction of GIS into projects and organizations are treated in reasonable detail.

1.2.3
How to use this book

This book has been written for the rapidly growing circle of professionals, academics and students from a wide variety of disciplines who have successfully completed their first experiences with Geographic Information Systems or computer-assisted cartography, and who wish to extend or update their knowledge about the application and implementation of time-integrative GIS.

As the readership comes from diverse scientific and professional backgrounds with dissimilar levels of knowledge and experience in GIS technology the chapters of the book are organised in a graded manner: Following a brief overview of the chapter's contents you will find a general introduction to basic terms and methods described. Advanced readers may skip those introductory parts and turn their attention to the specific problems treated in the subsequent sections of the text. The accompanying footnotes hold additional information and references as well as suggestions for further reading on the subjects and ideas discussed. With the help of the subject index at the end of the book more experienced readers and GIS users may look for specific information on certain aspects of temporal GIS.

1.2.4
Structure and organization of the book

The first chapter is conceived as introduction to the subjects, contents and structure of the book. Preliminary thoughts on computer based modelling of space and time are set out in chapter 2 by presenting some conceptual models on the interconnections of space and time as well as the most important fundamentals of Geographic Information Systems. Broader attention is given to the technical prerequisites as well as the theoretical concepts of the integration of time in GIS. As conclusion of a first thematic block chapter 3 discusses the temporal components of spatial data.

Chapter 4 is intensively concerned with the construction and implementation of a time-integrative GIS. Special attention is given to various aspects of data acquisition, data storage and data processing. While chapter 5 deals with the analysis of temporal data, chapter 6 discusses possibilities of visualizing spatio-temporal processes.

Chapter 7 presents practical approaches to build a historical GIS. Based on typical case studies taken from a wide variety of potential fields of application the theoretical concepts discussed in the previous chapters are illustrated.

The book is concluded by a summary as well as an outlook on current trends in commercial GIS packages (see chapter 8). An appendix contains a research bibliography, an annotated list of Internet resources as well as a CD-ROM with the examples presented in chapter 9.

In order to keep the book up-to-date and to provide additional resources the authors maintain a special website which can be found at http://www.time-gis.de.

1.3 Further reading

This book can only give a first introduction into the complex field of spatio-temporal GIS, which is closely interrelated with numerous adjacent fields and other disciplines. For obvious reasons the contents had to be concentrated on those principles and techniques which are absolutely necessary to enable the potential users to build their own spatio-temporal GIS applications with off-the-shelf software packages. Thus, neither all principles of geographic information theory nor all results of the vide variety of disciplines involved in GIS research could be included. Finally it was not possible to give comprehensive references and examples for all special fields of application in this volume. Therefore the following commented literature survey gives an overview of the relevant fields which should help the reader of this book to get access to relevant information necessary to widen the approach and perspective on spatio-temporal GIS. The list includes important research monographs as well as recently published hand- and textbooks. Please refer to our webpage at http://www.time-gis.de for an up-to-date version of the list.

1.3.1 Introductions in GIS and reference books

With the exploding number of GIS users the number of introductory texts on GIS has also seen a dramatic increase in the last decade. A neophyte's reading list might include the following titles: Bernhardsen T (1999) Geographic Information Systems. 2nd ed. Wiley, New York/Weinheim; Chrisman N (1997) Exploring Geographic Information Systems. John Wiley & Sons, New York; Clarke KC (1999) Getting started with Geographic Information Systems. 2nd ed. Prentice Hall, Upper Saddle River/NJ; Demers MN (2000) Fundamentals of Geographic Information Systems. 2nd ed. Wiley, New York/Weinheim; Heywood I, Cornelius S, Carver S (1998) An introduction to Geographical Information Systems. Longman, Harlow

The two weighty volumes of Longley PA, MF Goodchild, DJ Maguire, DW Rhind (eds) (1999) Geographical Information Systems, Vol 1: Principles and Technical Issues, Vol 2: Management Issues and Applications, 2nd edition, John Wiley & Sons, New York/Chichester comprise 72 contributions that discuss and explain the important issues and practices of geographic information systems. In addition to summarizing the state-of-the-art knowledge and best practices, it cross-references related themes and provides pointers to other texts, papers, and reports when appropriate. Another source of reference is provided by McDonnell R, Kemp K (1996) The International GIS Dictionary. Wiley, New York. This comprehensive dictionary provides an authoritative list of definitions for the most commonly used terms in geographic information science and neighbouring fields.

Readers interested in the historical development of GIS technology since the 1960ies may want to look at some classical texts like Burrough PA, McDonnell

RA (1998) Principles of Geographical Information Systems. Oxford University Press, Oxford; Marble DF, Peuquet DJ (eds) (1990) Introductory readings in Geographic Information Systems. Taylor & Francis, London; Star J, Estes JE (1990) Geographic Information Systems: an introduction. Prentice Hall, Englewood Cliffs; Taylor DRF (ed) (1991) Geographic Information Systems: the microcomputer and modern cartography (Modern Cartography 1). Pergamon, Oxford. The historical aspects are also treated in great detail by Foresman TW (ed) (1998) The history of Geographic Information Systems: perspectives from the pioneers (Prentice Hall Series in Geographic Information Science). Prentice Hall, Upper Saddle River.

German-speaking readers are recommended to have a look at Bartelme N (1995) Geoinformatik: Modelle, Strukturen, Funktionen. Springer, Berlin; Bill R (1999) Grundlagen der Geo-Informationssysteme, Bd. 1: Hardware, Software und Daten, Bd. 2: Analysen, Anwendungen und neue Entwicklungen. Wichmann, Heidelberg; Dickmann F (1999) Computerkartographie und GIS. Westermann, Braunschweig; Saurer H, Behr FJ (1997) Geographische Informationssysteme. Eine Einführung. Wissenschaftliche Buchgesellschaft, Darmstadt. In Dutch the reader edited by Hendriks P, Ottens H (eds) (1997): Geografische informatie systemen in ruimtelijk onderzoek. Van Gorkum, Assen is recommended.

1.3.2 Data capturing, data management, database technology and data storage

The specifics of geographical data bases are discussed by Adam NR, Gangopadhyay A (1997) Database issues in Geographic Information Systems (Kluwer International Series on Advances in Database Systems 6). Kluwer, Boston; Goodchild MF (ed) (1989) The accuracy of spatial databases. Taylor & Francis, London; Zeiler M (1999) Modeling our world: The ESRI guide to Geodatabase Design. ESRI Press, Redlands/CA. The latter places particular emphasis on object-oriented approaches and models.

Issues involved in GIS data capturing and conversion are covered by Hohl P (ed) (1998) GIS data conversion: strategies, techniques, and management. OnWord Press, Santa Fe/NM

Valuable insight into the management and processing of raster data is provided by Morain S (ed) et al. (1996) Raster imagery in Geographic Information Systems. OnWord Press, Santa Fe and in German by Göpfert W (1991) Raumbezogene Informationssysteme: Datenerfassung - Verarbeitung - Integration - Ausgabe auf der Grundlage digitaler Bild- und Kartenverarbeitung,, 2nd ed. Wichmann, Karlsruhe.

1.3.3
Land survey, cadastre, remote sensing and GPS

Besides information science and geography, geodesy and remote Sensing are other important pillars of GIS technology. The principles and contributions provided are explained e.g. by Hofmann-Wellenhof B, Lichtenegger H, Collins J (1997) Global positioning system. Theory and practice. 4th ed. Springer, Wien; Kennedy M (1996) The global positioning system and GIS: an introduction. Ann Arbor Press, Chelsea; Onsrud HJ, Cook DW (ed) (1990) Geographic and Land Information Systems for practicing surveyors: a compendium. ACSM, Bethesda; Teunissen PJG, Kleusberg A (eds) (1998) GPS for Geodesy. Springer, Berlin/Heidelberg/ New York; Campell JB (1996) Introduction to remote sensing. 2nd ed. Taylor & Francis, London.

From a German point of view, the following books may be added: Mansfeld W (1998) Satellitenortung und Navigation. Vieweg, Braunschweig/Wiesbaden; Albertz J (2000) Einführung in die Fernerkundung, 2nd ed. Wissenschaftliche Buchgesellschaft, Darmstadt.

1.3.4
Map projections and reference systems

Insight on the mathematical backgrounds, historical aspects and practical implications of different map projections and geodetic reference systems may be gained from Bugayevskiy LM, Snyder JP (1995) Map Projections. Taylor & Francis, London; Canters F (1989)The world in perspective: a directory of world map projections. Wiley, Chichester; Iliffe JC (1999) Datums and Map Projections. Whittles, Caithness; Maling DH (1993) Coordinate systems and map projections, 2ed Pergamon Press, Oxford; Pearson F (1990) Map Projections: Theory and Applications. CRC Press, Boca Raton; Synder JP (1993) Flattening the Earth : Two Thousand Years of Map Projections. University of Chicago Press, Chicago.

It might, however, be of crucial importance, to consult additional titles or documents issued by national surveying agencies in order to learn about the specific features of national map grids. In case of doubt additional advice may also be taken from technical white papers or manuals provided by the software manufacturer of your GIS package.

1.3.5
Data models, topology and spatial queries

Aspects of the creation of conceptual, logical and physical data models are treated in depths by Goodchild MF (ed) et al. (1996) GIS and environmental modeling: progress and research issues. GIS World, Fort Collins; Goodchild MF, Parks BO (1993) Environmental modeling with GIS. Oxford University Press, New York; Heuvelink GBM et al. (1998) Error propagation in environmental modelling with GIS (Research Monographs in GIS). Taylor & Francis, London.

Spatial query and reasoning techniques are discussed in Berry JK (1995) Spatial reasoning for effective GIS. GIS World, Fort Collins; Egenhofer MJ, Golledge RG (eds) (1998) Spatial and temporal reasoning in Geographic Information Systems (spatial information systems). Oxford University Press , Oxford; Ooi, BC (1991) Efficient Query Processing in Geographic Information Systems (Lecture Notes in Computer Science 471). Geometric algorithms are particularly emphasized by Kreveld M van (ed) et al. (1997) Algorithmic Foundations of Geographic Information Systems (Lecture Notes in Computer Science 1340). Springer, Berlin/Heidelberg/New York.

1.3.6 Spatial statistics, spatial analysis and spatial modelling

Spatial analysis and spatial statistics are a fundamental component of Geographic Information System. The techniques and algorithms applied are explained in a great number of books. Special attention on the linkages of GIS and spatial statistics is given by Chou YH (1996) Exploring Spatial Analysis in Geographic Information Systems. OnWord Press, Santa Fe; Fischer MM, Unwin D (ed.) (1996) Spatial analytical perspectives on GIS. Taylor & Francis, London; Fotheringham AS, Wegener M (eds) (2000) Spatial models and GIS: new potential and new models (GISdata). Taylor & Francis, London; Fotheringham S, Rogerson P (eds) (1994) Spatial analysis and GIS (technical issues in Geographic Information Systems). Taylor & Francis, London; Gaffney V, Stancic Z (1991) GIS approaches to regional analysis. Znanstveni Institut Filozofske Fakultete, Ljubljana; Longley P, Batty M (eds) (1996) Spatial analysis: modeling in a GIS environment. GeoInformation International, Cambridge; Molenaar M (1998) An introduction to the theory of spatial object modelling for GIS (Research Monographs in GIS). Taylor & Francis, London. Further insight is provided by Cressie, NA (1993) Statistics for Spatial Data. John Wiley & Sons, Chichester; Haining R (1990) Spatial Data Analysis in the Social and Environmental Sciences. Cambridge Unibversity Press, Cambridge.

1.3.7 Data warehousing and decision support systems

A growing number of projects and institutions use Geographic Information Systems for exploratory data analysis in large data holdings. The use of data mining techniques and the creation of decision support systems is described by Birkin M, Clarke G, Wilson A, Clarke M (1996) Intelligent GIS: location decisions and strategic planning. GeoInformation International, Cambridge; Leung Y (1997) Intelligent spatial decision support systems (Advances in Spatial Science). Springer, Berlin/Heidelberg/New York; Malczewski J (1999) GIS and multicriteria decision analysis. Wiley, New York/Weinheim; Thill JC (ed) (1999) Spatial multicriteria decision making and analysis: a Geographic Information Sciences Approach. Ashgate, Aldershot

1.3.8
Visualisation techniques, cartography and Web-GIS

The successful completion of a GIS project depends last but not least on the cartographic visualization of results. Standing in for the large number of books we recommend a closer look on the following volumes: Cartwright W, Peterson MP, Gartner G (eds) (1999) Multimedia cartography, with CD-ROM Springer, Berlin/Heidelberg/New York; MacEachren AM (1995) How maps work: representation, visualization and design. Guilford Press, New York; Unwin DJ, Hearnshaw HM (1994) Visualization in Geographical Information Systems. John Wiley & Sons, Chichester.

Two recently published titles deal principally with the dynamical visualization of space and time (in German): Buziek G, Dransch D, Rase WD (eds) (2000) Dynamische Visualisierung von Raum und Zeit: Grundlagen und Anwendungsbeispiele für kartographische Animation. Springer, Berlin/Heidelberg/New York; Dransch D (1997) Computer-Animation in der Kartographie. Theorie und Praxis. Springer, Berlin/Heidelberg/New York.

The Internet has far reaching consequences on the visualization and dissemination of spatial data. Written from a supplier perspective the following titles explain the underlying technologies and give practical advice on how to publish own GIS projects on the world wide web: Harder C (1998) Serving Maps on the Internet : Geographic Information on the World Wide Web. ESRI Press, Redlands; Plewe B (1997) GIS online: information retrieval, mapping and the internet. OnWord Press, Santa Fe. The list is complemented by the German-language publication by Gartner G (2000) Interaktive Karten im Internet. Ein Leitfaden. Springer, Berlin/Heidelberg/New York.

1.3.9
GIS project management and applications of GIS

The successful implementation of a GIS project requires a careful planning and depends on a number of crucial decisions. Typical pitfalls can be avoided by referring to the guidelines and procedures described by Aronoff S (1991) Geographic Information Systems: a management perspective. WDL, Ottawa; Huxhold WE, Levinsohn AG (1995) Managing Geographic Information System projects (Spatial Information Systems). Oxford University Press, New York/Oxford; Korte GB (1997) The GIS Book: understanding the value and implementation of Geographic Information Systems. 4th ed. OnWord Press, Santa Fe. Again, a book in German may be added: Behr FJ (1998) Strategisches GIS-Management. Wichmann, Heidelberg.

The potential applications of GIS technology are manifold. Besides the books dealing with general aspects of Geographic Information Systems, there is a growing number of titles targeted at specific user groups with special interests in GIS, listed in alphabetical order:

- *Archaeology:* Allen, KMS et al. (1990) Interpreting space: GIS and archaeology (Applications of Geographic Information Systems). Taylor & Francis, London; Lock GR, Stancic Z (ed) (1995) Archaeology and Geographical Information Systems: a European perspective. Taylor & Francis, London; Wescott K, Brandon RJ (2000) Practical applications of GIS for archaelogists: a predictive modeling kit (GIS Data Series). Taylor & Francis, London/New York
- *Architecture and Real Estate:* Castle GH (ed) (1998) GIS in real estate: integrating, analyzing, and presenting locational information. Appraisal Institute, Chicago; Hanna KC (2000) GIS for landscape architects with CDROM. ESRI Press, Redlands/CA
- *Biology:* McLaren SB, Braun JK (eds) (1993) GIS Applications in mammalogy. Oklahoma Museum of Natural History, Norman; Savitsky BG, Thomas E, Lacher J (eds) (1998) GIS methodologies for developing conservation strategies: tropical forest recovery and wildlife management in Costa Rica (Biology and Resource management). Columbia University Press, New York; Wadsworth R, Treweek J (1999) GIS for ecology: an introduction. Addison Wesley Longman, Harlow
- *Energy and Utilities:* Harder C (1999) Enterprise GIS for energy companies with CDROM. ESRI Press, Redlands/CA
- *Environmental Monitoring and Physical Geography:* Bähr HP, Vögtle T (1999) GIS for environmental monitoring. Schweizerbart, Stuttgart; Carrara A, Guzzetti F (eds) (1995) Geographical Information Systems in assessing natural hazards (Advances in Natural and Technological Hazards Research 5) Kluwer Academic, Dordrecht; Douglas WJ (1995) Environmental GIS: applications to industrial facilities. Lewis Publishers, Boca Raton/London
- *Health Care:* Gatrell A (1998) GIS and health (Gisdata 6). Taylor & Francis, London/Philadelphia; Lang L (2000) GIS for health organizations with CDROM. ESRI Press, Redlands
- *Hydrology:* Abbott MB (1991) Hydroinformatics: information technology and the aquatic environment. Aldershot, Avebury Technical; Gurnell AM, Montgomery DR (eds) (2000) Hydrological applications of GIS (advances in hydrological processes). Wiley, Chichester/Weinheim; Singh VP, Fiorentino M (ed) (1996) Geographical Information Systems in hydrology (Water Science and Technology Library 26). Kluwer, Dordrecht; Wright DJ, Bartlett DJ (eds) (2000) Marine and coastal Geographical Information Systems (Research Monographs in Geographic Information Systems.). Taylor & Francis, London
- *Marketing:* Grimshaw DJ (2000) Bringing Geographical Information Systems into business. 2nd ed. Wiley, New York/Weinheim; Harder C (1997) ArcView GIS means business. ESRI Press, Redlands; Leiberich P (1997) Business Mapping im Marketing. Wichmann, Heidelberg (in German); Longley P, Clarke G (ed) (1995) GIS for business and service planning. GeoInformation International, Cambridge
- *Resource Management:* Belward AS, Valenzuela CR (eds) (1991) Remote Sensing and Geographical Information Systems for Resource Management in

Developing Countries (Eurocourses: Remote Sensing 1). Kluwer Academic, Dordrecht; Heit M et al. (eds) (1996) GIS applications in natural resources 2. GIS World, Fort Collins; Lang L (1998) Managing natural resources with GIS. ESRI Press, Redlands
- *Sociology:* Martin D (1996) Geographic Information Systems: socioeconmic applications. Routledge, London
- *Transportation and Logistics:* Lang L (1999) Transportation GIS. ESRI Press, Redlands/CA
- *Urban and Regional Planning:* Greene RW (2000) GIS and public policy with CDROM. ESRI Press, Redlands; Kilchenmann A, Schwarz von Raumer HG (1999) GIS in der Stadtentwicklung. Springer, Berlin/Heidelberg/New York (in German); Stillwell JCH (ed) et al. (1999) Geographical Information and planning (Advances in Spatial Science). Springer, Berlin/Heidelberg/New York

1.3.10 Manuals and programming language handbooks

For a number of reasons users are often not very satisfied with the manuals that come with an of-the-shelf GIS package. The remaining gaps are filled with instructive titles either giving a general introduction or dealing with special aspects of the software.

The software packages provided by ESRI are covered by a whole string of titles, often published by the company itself. Editors of ESRI Press (1999) Getting to know ArcView GIS. 3rd ed. ESRI Press, Redlands; Madej E (2000) Cartographic design using ArcView GIS. OnWord Press, Albany; Verbyla DL, Chang, KT (1997) Processing digital images in Geographic Information Systems: a tutorial featuring Arcview and Arc/Info. OnWord Press, Santa Fe/NM; Liebig W (1999) Desktop-GIS mit ArcView GIS. Leitfaden für Anwender. Wichmann, Heidelberg (in German). Besides these introductory volumes there is a number of books addressed to advances users that want to extend the software's capabilities by using the built-in programming language (Avenue) or by working with the mapping component (MapObjects) that adds GIS functions to any software project: Ormsby T (1999) Extending ArcView GIS: with network analyst, spatial analyst and 3D-analyst. ESRI Press, Redlands; Razavi AH, Warwick V, Alexander J (2000) ArcView GIS/Avenue Programmer's Reference. 3rd ed. OnWord Press, Albany; Hartman R (1997) Focus on GIS component software: featuring ESRI's mapobjects. OnWord Press, Santa Monica.

The products of MapInfo Corp. are dealt with by Daniel L, Whitener A, Loree P (1999) Inside MapInfo professional 2nd ed. OnWord Press, Albany; Whitener A, Davis J (1998) Minding your business with MapInfo. OnWord Press, Santa Fe; Whitener A, Ryker B (1997) Mapbasic developer's guide. OnWord Press, Santa Fe.

Regardless of a smaller market-share other software products like Geomedia and Autodesk World are also covered: Harmon D, Limp WF (1998) Inside Geomedia. OnWord Press, Santa Fe; Vance D et al. (1998) Inside Autodesk World. OnWord Press, Santa Fe/NM.

The websites of the major GIS manufacturers offer an abundance of information sources, such as white papers, manuals, code examples and software for free download.

1.3.11 Philosophical, social, political and legal aspects of GIS

Besides the technical aspects, the use of GIS implies a number of social, ethical and legal issues. These questions are raised by Aldenderfer MS, Maschner HDG (eds) (1996) Anthropology, space, and Geographic Information Systems (spatial information series). Oxford University Press, New York; Cho G (1998) Geographic Information Systems and the law: mapping the legal frontiers. John Wiley & Sons, Chichester/New York; Curry MR (1998) Digital places: living with Geographic Information Technologies. Routledge, London; Eagles M (ed) (1995) Spatial and contextual models in political research. Taylor & Francis, London; Masser I (1998) Governments and Geographic Information. Taylor & Francis, London; Pickles J (ed) (1995) Ground truth: the social implications of Geographic Information Systems (Mappings). Guilford Press, New York.

2 Conceptualising real world entities in spatio-temporal GIS

„Geography differs from geometry because in geography, space is indivisibly coupled with time"
(Don Parkes & Nigel Thrift 1980)

The modelling of real world phenomena is one of the central and most difficult duties in building a GIS. After reflecting some basic thoughts on the perception of space and time in geography and the social sciences, chapter 2 deals with various aspects related to the conceptualisation and modelling of real world entities in GIS.

2.1 The geographer's perspective on space and time

Describing an ordinary scene on the beach as an example, Haggett (1979) asked, "what makes such a scene of special interest to geographers?" The answer given in his classical introductory chapter "On the Beach", was since then used to explain the geographer's perception of the world to generations of geography students. Basic principles and problems connected with the relationship between our understanding of the world and the framework of implementing a research process targeted on the human environment can be derived from this example. It will be used to explain some of the basic concepts necessary for the development of a spatio-temporal GIS.

For geographers the geographical distribution of objects is of main concern. Looking at a specific part of the beach as an object of research, we may observe people distributed all over the beach. Each person can be identified by the place he or she is located at the moment of observation. Space, location and place are important terms to describe where observations are made:

- *Space* is usually utilized as an abstract term to describe the expansion of the earth's surface. In general, the term is not precisely determined. Space is widely used as subject or means of explanation in geography, and intensely interrelated with the definition of geography as a science itself.[4]
- *Location* denotes a specific part of the earth's surface, distinguishable in one way or the other from the surrounding space. Usually not precisely defined, the term may be used to refer to particular positions on the earth's surface.

[4] For further discussion and references on this essential discussion in geography see Johnston et al. (1996:573-575).

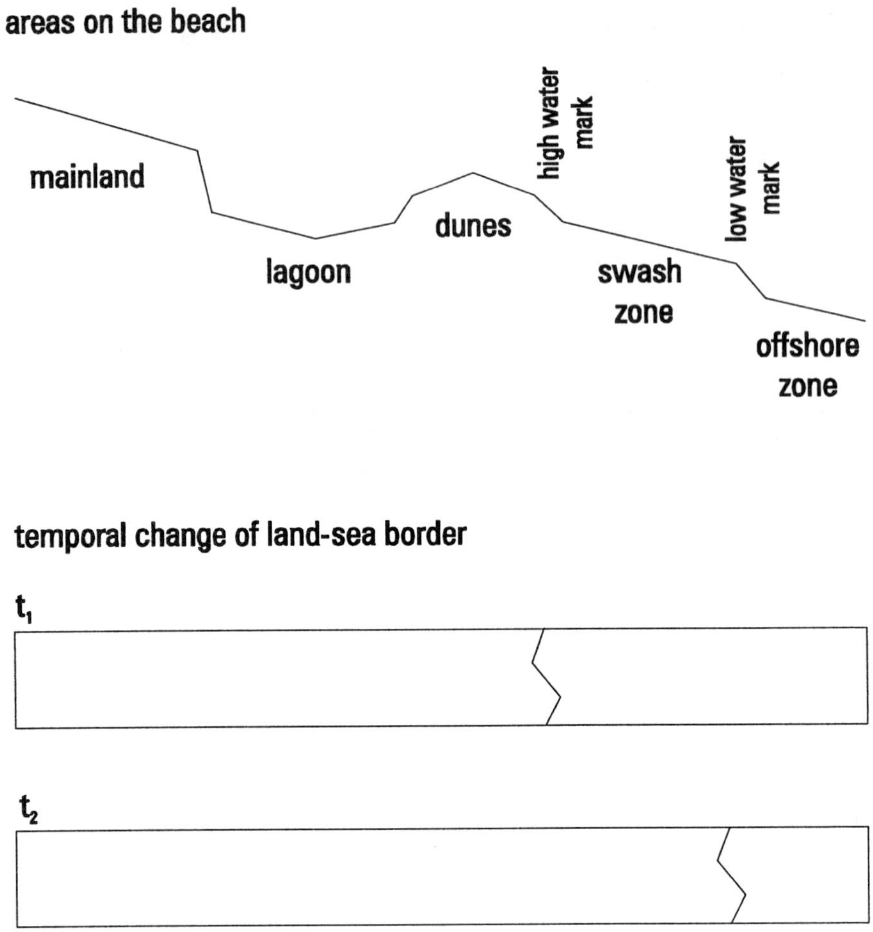

Fig. 2.1 Haggett's "On the beach" (adopted from Haggett 1979:6)

- *Place* is used for defining a particular position on the earth's surface which can be given an unmistakable designation. This could be a place name, an address or a geographical coordinate, which is the most common way to address the position of places in geography.

It is obvious that such abstract definitions are not sufficient to describe the geographical distribution of observations in a scientific sense. More sophisticated techniques are necessary to locate the exact position of an event or observation on the earth's surface. Geographical reference systems like longitude and latitude are used for this purpose. Also measures of distance like meter or kilometre are used to define the interrelation of objects. Besides this absolute space, measured in meter and kilometre or longitude and latitude, in human geography also non geometric concepts of space are used.[5]

Because of the large number of people on the beach, the pure information of the particular position of each person gives us no understanding how and why this particular distribution on the beach developed. To understand the underlying processes which cause the observed spatial distribution of people on the beach we have to look at the beach as a whole. The actual location of persons can be aggregated and described in terms of spatial patterns of particular distributions. With this aggregated information we may derive more meaningful statements, e.g. that there are areas on the beach where more people can be observed than in other areas.

This leads to the question whether these patterns of distribution are perhaps caused by other characteristics or features of this areas. Therefore we can try to describe the surface of the beach in a geographical way by distinguishing several areas with different qualities. Such areas may be for instance the offshore zone, the swash zone, the dunes and the mainland (see figure 2.1). By comparing the areas with highest density values of people, we may now try to figure out, what factors caused the preference of this respective area by the visitors. The definition used for this purpose and the boundaries resulting from them are neither initially given or inherent nor can those boundaries be seen in the real world. By using expressions like swash zone we are interpreting the differentiation of the surface by applying theoretical insights in the origin of the beach. We also make assumptions on the causal relations between the spatial distribution of people on the beach and the attributes of certain areas.

The definition of areas which subdivide the beach into zones with homogeneous properties will not always lead to spatially exact boundaries. Sometimes continuous phenomena of the real world can only be represented by discrete areas if classification is applied. The example of the beach shows, that the differentiation between land and see was made using an additional area category, the splash zone. The splash zone in this example is defined as the area between the highest and lowest sea level. We have to consider, that in reality there is no boundary on the beach indicating the splash zone, it is our conceptualising of the beach-sea system, based on theoretical knowledge. In this case we use our knowledge that there is a tide and resulting changes of the sea level, so that the border between beach and sea will vary over time.

However, this point of view is closely related to scale. Using a smaller scale, a smaller number of zones are distinguished and the zone of transition of land and sea may be reduced to a border line. Using a larger scale numerous additional zones may be identified and the zone of transition may be reproduced more exactly.

If we once again look at the spatial distribution of bathers in the different zones of the beach, we may eventually find out, that the initial assumption was wrong: There is no correlation between the distribution pattern and the beach zones. The divergence from the assumed interrelation may result from a misconception: We

[5] Cf. Gatrell (1991) for examples.

dealt with the visitors as a homogeneous group and assumed that everybody had similar preferences of finding a suitable location for staying on the beach. In order to exclude this source of error, we need to collect additional information on the visitors, such as age, sex et cetera. In the next step we need to group the visitors according to the attributes surveyed in order to analyse the distribution patterns of the groups. We may find out that the distribution patterns of some groups match perfectly with the different beach zones, while the location of other groups can not be explained sufficiently. As in reality spatial distributions can hardly ever be explained with a single cause, we need additional information on the spatial differentiations of the beach.

If there was an additional attraction such as an ice-cream booth, the surrounding areas of the beach would be preferred by certain groups of visitors. Such causal explanations and corresponding methods of analysis are frequently used in geography in order to explain spatial patterns.

The location of particular visitors on the beach and the resulting spatial patterns are also subject to temporal changes. Differences in spatial patterns can be observed between different times of the day, working days and weekends, summer and winter et cetera. In this case we must again make assumptions on the causal connections, in order to select suitable moments of observation. If there is a video camera at our disposal, we may continually tape the activities on the beach. Otherwise we may frequently visit the beach to map the distribution of visitors. If we do not only observe one point in time, we need to adjust our zoning of the beach to the changing conditions, e.g. of tides and seasons. Obviously, the complexity of the information given increases considerably. We end up with a complex spatio-temporal model of the real world which includes assumptions and theoretical concepts on causal interdependencies to be studied.

In the case of our book we need to answer the question, how to transfer such a conceptual model of reality into a GIS. In order to achieve this, our model must be composed of structural elements, which can be translated into the formal structures of a database and a GIS software. While this problem will be discussed in chapter 2.2, section 2.3 describes the different data structures in a GIS. However, the conceptualisation of the real world into spatial objects is subject to numerous methodological problems; some of them will be explained in chapter 2.4.

2.2 Modelling the real world in Geographic Information Systems

The modelling of real world phenomena ranks among the central and most difficult tasks of building a GIS. In general, science uses models to structure and reduce the abundance of information on the environment (Hake & Grünreich 1994:14). Stachowiak (1965:438) mentions the following characteristics of a model:

- *Illustration:* Models may represent either natural or artificial objects, that can be models themselves.
- *Simplification:* The model building process intends a reduction of complexity by identifying relevant and irrelevant features.
- *Subjective pragmatism:* The reduction and simplification of the real world underlies subjective decisions. Depending on different persons, situations and subjects a system may be transferred into differing models.

However, in empirical sciences such as geography, this general conception of a model must be restricted to the fact, that the quality of the model must be verifiable by empirical findings (Hake & Grünreich 1994:14). For the application inside a GIS the definition must be extended to the fact, that a model depicts a part of the real world which can be represented by data. Goodchild for example defines a data model as „a limited representation of reality, constrained by the finite, discrete nature of computing devices." (Goodchild 1992:401). This definition addresses the fact, that the complexity of the real world must be reduced, so that it can be stored in a GIS. One important source of reduced complexity is due to the constraints imposed by the use of information technology. Computers and software use formalized rules to process data, so the data itself has to be in a finite and discrete form (Frank & Egenhofer 1992:976; Bittner & Frank 1997:12; Herring 1992:445).

In connection with a reduced complexity the modelling of the real world implies the loss of a lot of information and meaning during the process. We have to choose which information is included and which can be neglected. The decision which information to be included is based on our judgement whether a piece of information is important or not. While doing this we execute a distinct reinterpretation of the real world. Therefore, a model can never be valid for all purposes. Another source of falsification within the modelling process is connected with the way in which we use finite and discrete information to represent infinite and continual phenomena of the real world. The choices made are clearly driven by the objectives of the model and the kind of database technology we use.

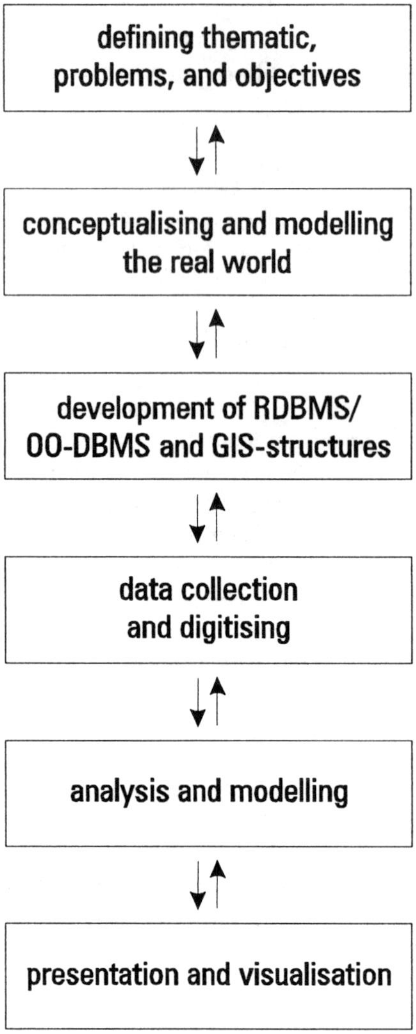

Fig. 2.2 Steps in the development of a GIS (adopted from Hendrik & Ottens 1997:19)

2.2.1 Steps in building a GIS

The reduction of complexity of the real world during the modelling of a GIS may lead to different results. The modelling process itself is not only related to the real world but also to the constraints imposed by GIS technology. Thus, the outcome of the implementation process may vary considerably, and a lot of mistakes can be made. The representation of the real world may overlook important phenomena, or the modelling may not represent the real world according to the objectives of the GIS. Another type of problems can arise from technical misconceptions, so that data in the GIS may not be processed, analysed and presented as desired. It is therefore highly recommended not to build a GIS based on an ad hoc analysis of requirements. There are several stages of planning, which may help to prevent mistakes up to the complete failure of the GIS project (see figure 2.2).

First the general subjects of the GIS have to be defined. The problems which shall be solved with the GIS have to be identified and the objectives of the GIS must be stated. The problems under investigation determine the necessary analysis and presentation tools. These depend e.g. on the data structure which needs to be designed before the analysis and presentation can be applied. The objectives are also important for the outcome of the GIS. If the GIS shall foster research, a versatile and flexible design is recommended. If the GIS shall improve the storage and query of spatio-temporal data, the system performance is of main concern. If the GIS shall provide mostly standardized information services the system integration and availability of suitable user interfaces is important. All objectives given here as an example will have an effect on the selection of a data structure and the evaluation of an appropriate software package. Therefore this step should be discussed in detail beforehand. Goals not clearly defined will threaten the whole building process. The decisions which have to be made right from the beginning are related to the environment in which the GIS will be used, the resources and expertise which are available, and the time in which the GIS has to be developed. These aspects of the implementation process are discussed in greater detail in chapter 4.

The second step in building a GIS is the conceptualising of the real world. For this stage a great deal of experience and theoretical knowledge on the nature of real world phenomena is necessary. The conceptualisation depends on methodological considerations, which are discussed in the following chapter. Following the decisions on how and by which entities and data the real world should be represented in the GIS, the data structures that store the relevant information must be developed. We have to differentiate between the storage of the spatio-temporal attribute data on the one hand and the related geometry data on the other hand. The data structure will depend on the database and GIS technology available and vice versa. The decisions on data structure and the related database management and GIS technology are again interdependent with the demands for analysis and presentation.

During the next phase the data must be collected and the database must be built. Furthermore the geometry objects must be digitised or imported into the GIS. This is a critical point, where all decisions made before must be re-evaluated carefully. The data collected for a GIS is often the most expensive part of the building process as it might be used long after the hard- and software ceased to exist.

The subsequent analysis and modelling is the most interesting part of the building process. At that point mistakes or misconstructions made before would have crucial impacts. Finally the results may be visualised and presented. Both steps will be discussed in chapters 5 and 6.

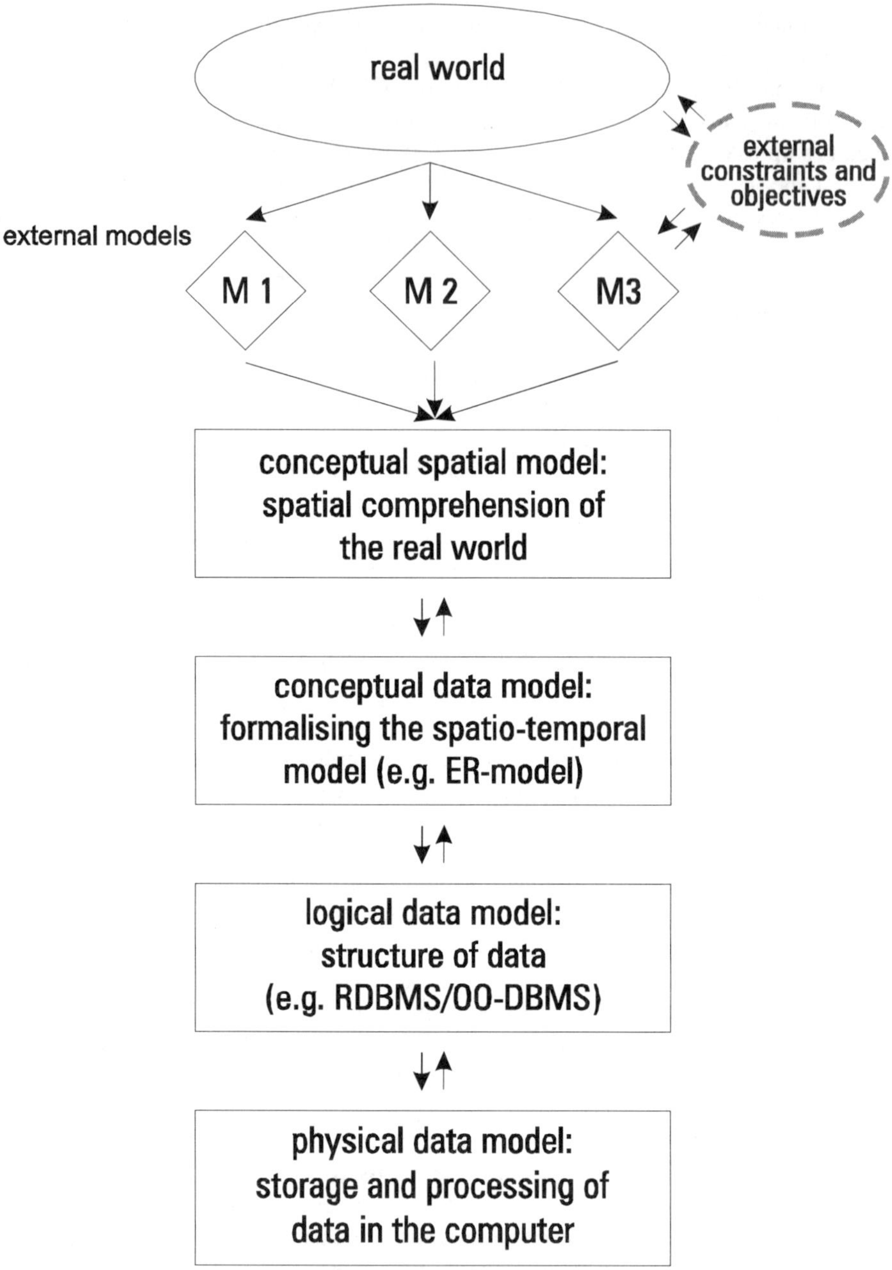

Fig. 2.3 Conceptualising the real world in GIS (adopted from Hendrik & Ottens 1997:5)

This linear succession of stages in the building of a GIS does not necessarily correspond with the respective stage during the planning process. As mentioned above all of the steps are interrelated with each other. The planning process must consider these effects in advance and eventually re-evaluate the planned procedure. Before the building of the GIS starts, necessary adaptations can prevent a unsuccessful implementation. During the implementation (see chapter 4) the evaluation of the completed GIS and efforts for the update and maintenance will be the final stages.

2.2.2 Stages of abstraction in modelling a GIS

The process of representing the real world in a GIS may be separated into different stages, starting with the conceptualising of the real world (see figure 2.3, Peuquet 1990:252).[6] As discussed before, the real world could be understood as a spatio-temporal continuum. Our understanding of the real world and therefore our conceptualising of the real world depends on our view of the world, our experience, our theoretical knowledge, and last but not least our interest and objective. These influences on our view of the world are commonly addressed as external models. They can be clearly defined and formalised models, maybe representing research results or hypothesises. In practice they are often undeclared assumptions or prejudices. At this stage the interests and goals of the user are of main concern (Laurini & Thompson 1992: 357).

The conceptualisation aims at giving a formalised description of the comprehension of the real world. This formalisation should be as clear and unmistakable as possible. At the other hand the conceptualisation should not rely on a specific way of modelling or building the GIS at this stage. The level of content and the level of modelling should be separated. This is of great practical importance, because of the different ways to build a spatio-temporal GIS, depending on a changing technology.

Because this type of modelling deals with meaning, notion and terms, it can also be addressed as *semantic modelling* of the reality (Scheer 1995:175). The conceptual spatial model can rarely be applied directly in order to store data in an information system. In a second stage one has to build clearly defined elements of the real world. Such elements are commonly called entities (see chapter 2.2). The relationship of such formal elements can for instance be depicted in an entity-relationship model (ER-model).

The information assigned to such final and discrete objects can be stored in an information system. This stage is called the logical data model. On this level, the information has no inherent structure. The most simple way to store the data is an unordered list. However, the data must not only be stored but be accessible later on. Thus, in terms of performance such an approach is not appropriate. There are

[6] This schema for the conceptualising of the real world in digital information systems originates from the American National Standards Institute, Standards Planning and Requirements Committee (ANSI-SPARC) and is now a common standard (Laurini & Thompson 1992:351).

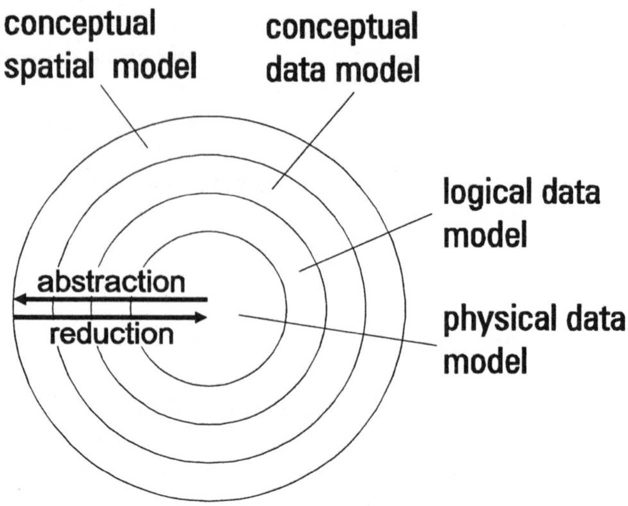

Fig. 2.4 Stages of data modelling (adopted from Molenaar 1997:21)

different ways in which data could be structured logically, depending on the database technology being used (Laurini & Thompson 1992:361). The transfer from the conceptual data model to the logical data model is an intellectual process. The formal structure given in the conceptual model can guide this process. Often the ER-model can be used as a blueprint for the development of tables in a relational database management system. However, this is not necessarily the optimum: criteria for finding the best data structure for a project may also be derived from the project requirements. Different preferences such as performance, redundancy free storage etc. determine the natural constraints for this judgment. There are always different solutions available from which the best must be evaluated (Scheer 1995:46; Schneider 1995:15).

The differentiation between the conceptual and the logical level of the modelling process should not be seen as an end in itself. The conceptual level encourages the aim of the modelling, the interpretation of the real world and the knowledge used to define entities which are utilized to reduce the complexity to a level which can be stored in an database. The logical level in contrast focuses the interest on the computing methods used to facilitate these aims (David et al. 1993:266-267).

The physical model, which marks the end of the modelling process, may not be accessed by the user. It describes the physical storage of the data on the storage media (Laurini & Thompson 1992:362).[7]

---

[7] Cf. Molenaar (1998:7). Peuquet (1984:69) uses the terms of reality, data model, data structure and file structure for the different levels of abstraction.

2.2.3
Conceptualising the real world in Geographic Information Systems

The preceding section examined the steps from the conceptualising of the real world to the logical data model. This process can be depicted as concentric rings with a decreasing stage of abstraction and an increasing stage of reduction (see figure 2.4). This chapter deals with the conceptualisation of spatio-temporal processes in a GIS. [8] The intention is to give a brief introduction on this complex issue for readers who – at the moment – are not familiar with the wide variety of methodological discussions and practical approaches.

2.2.3.1
From entities to objects

The conceptualisation of a phenomenon of the real world must reduce the complexity, so that one distinct element or process can be distinguished. As explained before in a more generalized view, an understanding of the spatio-temporal dimension of the phenomena of the real world is used to derive a formalised spatio-temporal view. Such an element, or process may be conceptionalised by different spatio-temporal representations (see figure 2.5). A process might for example be represented by points, indicating the location of certain events. Another view on the process conceptualises the occurring of the events according the their spatio-temporal diffusion using lines. Another view might utilise regions to sum the number of events in a given area or symbolise the density of such events by applying contour lines.

Entities and objects. The terminology used for GIS is heterogeneous. The names used to describe elements of the real world and elements of the GIS are often misleading. Sometimes "feature" and "object" are used to differentiate between the real world sphere and the data used in a GIS (ANSI 1997a:7), other sources use "geoobject" for elements of the real world and the GIS without further differentiation and "feature" for the corresponding data stored in a database (Lange 1997:15ff). The inconsistency of the terminology becomes even worse when time is added to the dimensions of the GIS. Langran (1992:48ff) distinguishes between "objects" and "features". Objects are defined by their location and are related to attribute data which may vary. Features are defined by their content, but form and position may change through time. Voigtman (1997:50f) in contrast uses those terms with the opposite meaning.

In this book the term "entity" is applied to elements of the real world, which can be distinguished from other elements, using distinct characteristics of the element. According to Laurini & Thompson (1992:362) "an entity is a person, place, thing or event. The Mr Green's parcel, the building 2314, the car number 34HP89, the flight 123, are entities in the real world." Entities may also consist of

[8] Cf. Mark (1999), Martin (1999), Raper (1999) and Smyth (1999) on the ontological and epistemological problems of spatio-temporal representation.

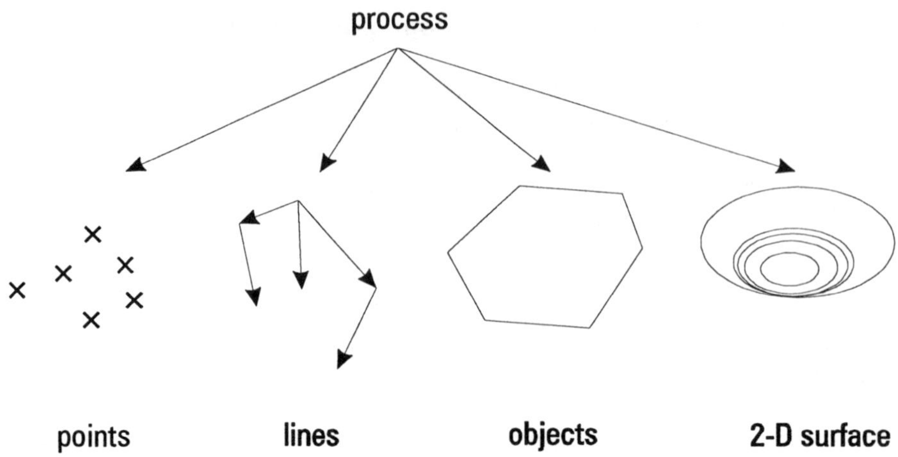

Fig. 2.5 Different spatial conceptualisations of a process (adopted from Raper 1999:75)

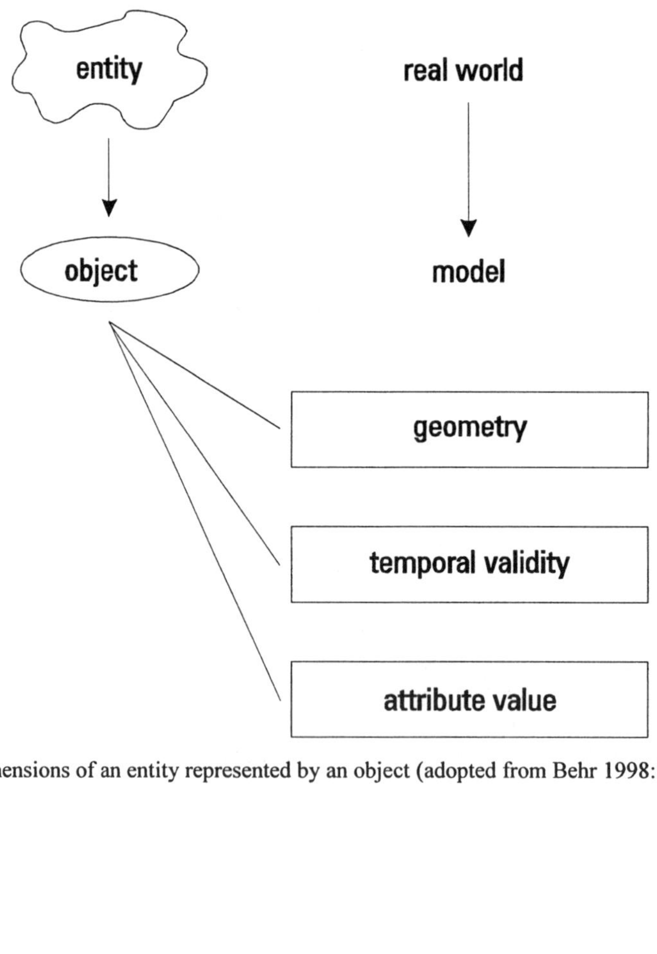

Fig. 2.6 Dimensions of an entity represented by an object (adopted from Behr 1998:131)

other entities of a different type. Entities do not only represent elements physically present in the real world, they also abstract and represent conceptualisations such as events or processes.

The spatio-temporal representation of an entity in a GIS is an object. Objects consist of a spatial dimension, a temporal dimension and additional attribute data (see figure 2.6). The spatial dimension of the object can be modelled in different ways. The different representations of space in a GIS are discussed in detail in section 2.3, the representation of the temporal dimension in chapter 3.

Objects in a GIS must be structured, so that the GIS can address them. Two conceptual strategies are available: the organisation of the information, whether on the level of entities or objects, can be made by using layers or an ER-model.

Layer-GIS-Model. The layer GIS model is one of the basic concepts of GIS, applied in most desktop GIS packages. Using layers, information can be structured logically according to different criteria. Entities or objects may be arranged in a way that each layer consists of only one kind of information (see figure 2.7). A subdivision in e.g. theme, time or storey could be used to categorize objects. The layer concept is similar with the overlay of transparencies used in traditional cartography, where one layer is used for every type of information on the map (Bartelme 1995:42).

The spatial representation of the respective objects is given by their geometrical extent. This means that all objects are grouped according to additional properties, while the geometry is used as a reference. In contrast to the representation of space, the representation of time is only possible related to the spatial objects. In the case of a layer concept this means that the objects are grouped according to related temporal information.

ER-model/Object-relational-Model.[9] In an ER-model the relationship among entities is structured according to the hierarchy of the entities derived from related attribute information (see figure 2.8). The object-relational-model groups similar objects to classes, which can be part of another class in a hierarchically structured relationship (Bill & Fritsch 1994:241-243). Figure 2.9 shows the structure of an object-relational data model. Each object consists of the attribute data and the related object methods or functions (Schader & Rundshagen 1994:15, Worboys 1999:378-9). The main advantage of object-relational-approaches lies in a consistent data structure, the prevention of redundant data storage, the convenient transformation from the conceptual to the logical data model, and the operational advantages of object-oriented databases.

Similar to the ER-model, the object-relational-model may be applied to different purposes, e.g. for structuring objects in a GIS. "The OO approach is in use both as a method of semantic data modeling and as a model of data handled by object-oriented programming and database management systems." (Worboys 1999:378).

[9] Cf. e.g. Bédard (1999) on ER-models and Worboys (1999) on object-oriented models.

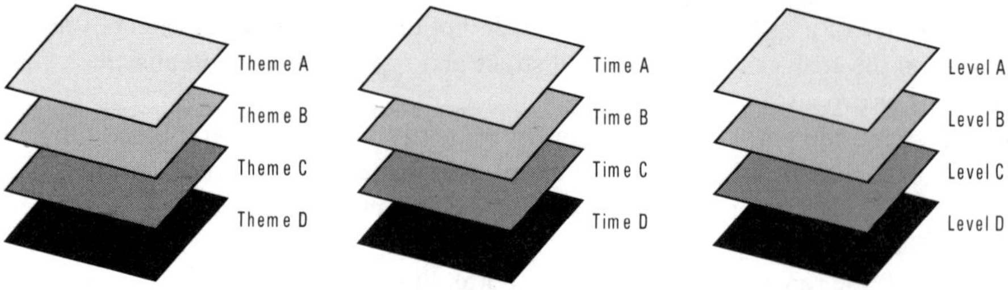

Fig. 2.7 Layer-GIS-model (adopted from Laurini & Thompson 1992:7; Bill & Fritsch 1994:17)

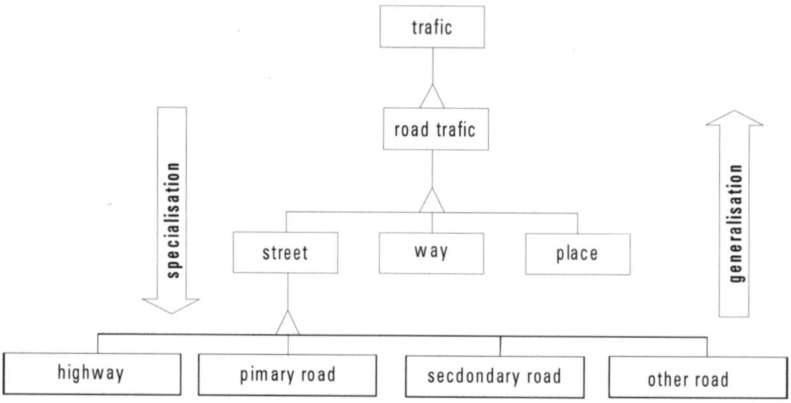

Fig. 2.8 ER-model (adopted from Hellwig 1999:39)

In figure 2.10 the relationship between spatial objects is depicted. Compared to the layer concept the advantage of the object-relational-concept is, that an object can be part of different classes. Different perspectives on one object can be realized by including the object in different classes. Compared to the layer model the object-relational model can directly cover spatial relations between objects and classes of objects. Although object-oriented approaches for GIS are widely discussed and the advantages are widely acknowledged, most desktop GIS packages do not support object relational data structures.

2.2.4 Geo-coding

After defining and structuring spatio-temporal objects the objects must be linked to the corresponding attribute information. The allocation of attribute data to the respective spatial representation is usually called geo-coding. In the most simple case of spatio-temporal data the geographic representation is given by a direct relation of position to attribute data (see figure 2.11). Such a data structure is for

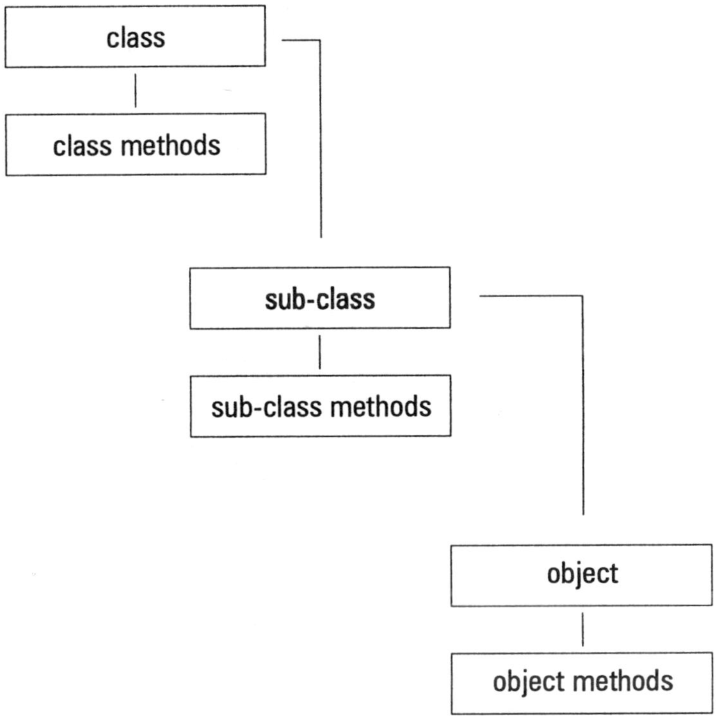

Fig. 2.9 Object-relational-model (adopted fromMolenaar 1997:38,40)

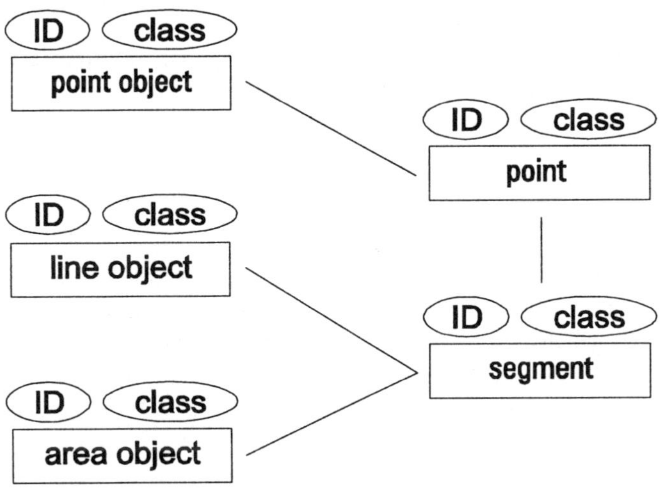

Fig. 2.10 ER-model for vector objects in an OO-environment (adopted from Molenaar 1997:36)

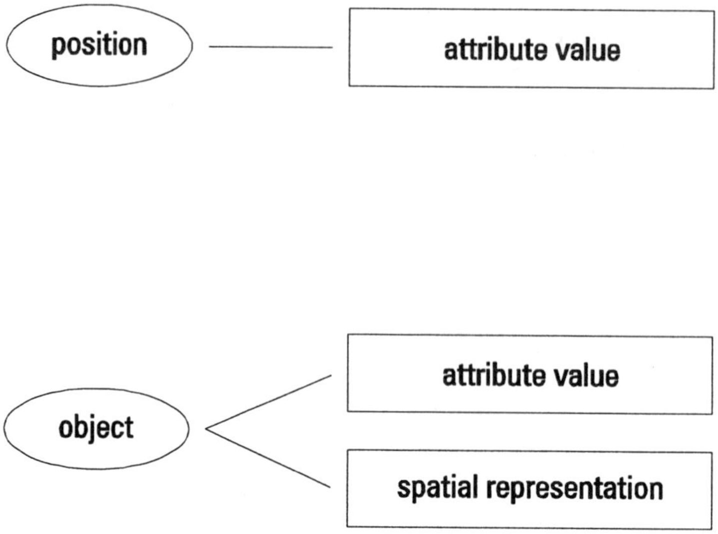

Fig. 2.11 Two basic structures for the representation of spatio-temporal data (adopted from Molenaar 1997:23)

example the result of fieldwork, where coordinates and attribute values were stored in a table for later use. In another case spatial objects are defined by a spatial representation and the related geo-coded attribute data. Frequently this is the case when the spatial representation and the attribute values are taken from different sources.

Geo-coding is a basic concept in GIS, linking spatial representations with attribute information. However, in practice the identification of objects which correspond to the data and vice versa is a difficult task.

2.2.5 Geo-referencing

The spatial information of a geo-object is not given a priori. The locations on the three dimensional surface of the earth have to be transformed, in order to be stored in a data format which is suitable for the visualisation on a two dimensional map or computer screen. This process, called geo-referencing, is also a basic concept in GIS. Map projections and reference systems are needed to relate the spatial representation of an object to the earth's surface.

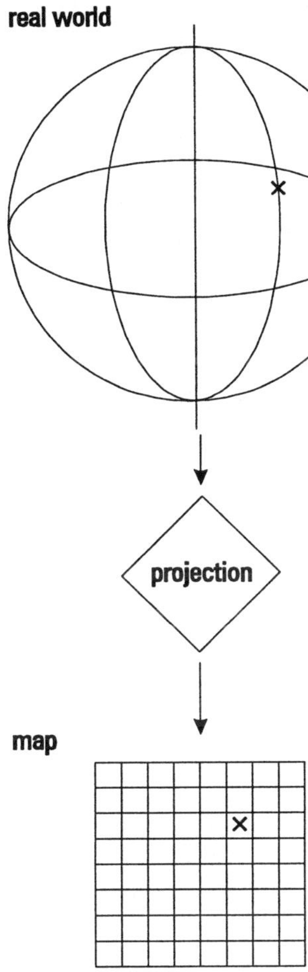

Fig. 2.12 Map projection

2.2.5.1 Map projections

In the three-dimensional system of the earth's surface each position can be given by simply using the angle of longitude and latitude. For the two dimensions of a map, longitude and latitude have to be transferred (see figure 2.12). Map projections provide rules how a pair of three-dimensional coordinates can be mapped on a two dimensional map. Map projections are important whenever we need to digitise paper maps for the use in a GIS. Data mapped or stored in different map projections may not be used simultaneously and must be transformed. [10]

[10] Cf. e.g. Bugayevskiy & Snyder (1995) and Snyder (1997) on map projections

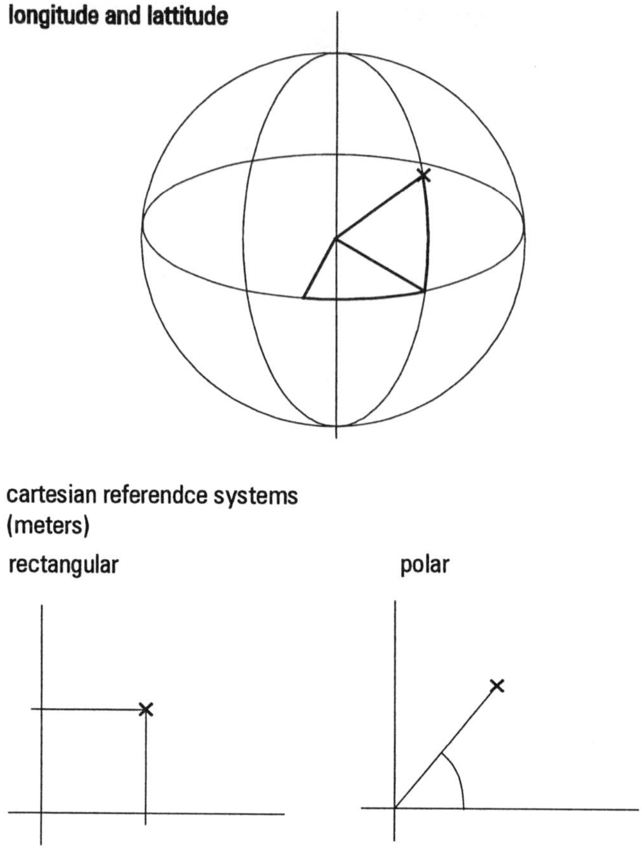

Fig. 2.13 Reference systems

2.2.5.2
Reference systems

The reference systems used in GIS can be differentiated into geographic and Cartesian reference systems (see figure 2.13). Geographic reference systems use longitude and latitude as measures. Cartesian reference systems use a planar rectangular coördinate system, where a point can be identified by rectangular or polar values. In the Cartesian reference systems like UTM or Gauß-Krüger distances may be measured directly. Because of growing distortions produced by the projection applied to generate a Cartesian reference system such systems are used for large and medium scale levels only. [11]

---

[11] Cf. e.g. Seeger 1999 on spatial referencing and coordinate systems.

2.3 The representation of space in GIS

Space is commonly represented in two different ways. Figure 2.14 shows that any given object can spatially be represented by a raster or a vector data model. Currently a lot of GIS packages support the vector and the raster data model. In most cases both data models are not directly linked with each other. They make use of the fact, that each vector object can easily be transformed into a raster data format.

Each geographically defined point in the real world can be part of different spatial data models in a GIS (see figure 2.15). Besides the storage in a raster data model and a digital elevation model, a point may be part of a point, line or region object in a vector GIS.

2.3.1 Vector data models

With a vector data structure, entities of the real world must be modelled reducing the spatial complexity. Figure 2.16 depicts, that e.g. a river as a real world entity can be modelled spatially using different spatial objects. The conceptualisation of a spatial model of an entity depends on the requirements of the system. The different line objects give an example of different stages of generalization. With line objects the river is symbolized in its linear nature. The fact that in the real world the river has an area is neglected. Thus, only a region object allows the representation of not only the linear character of the river but also its spatial extension.[12]

2.3.1.1 Vector data objects

The vector data model uses three types of objects (see figure 2.17). An overview of the properties of the three vector data types in terms of spatial queries and operations is given in chapter 5.

- *Point objects* are clearly defined by one pair of x-y-coordinates. Each point object can be used to geo-code additional attribute information.
  Point objects are in fact scale dependent abstractions because no entity present in the physical sphere can be completely described by a point coordinate without a spatial extent. Point objects are used in all cases, where the extent of an entity may be neglected. This is true for entities with a relatively small spatial extent, such as trees. Such objects are too small to be represented by their spatial extent in scales usually applied in GIS. The scale dependency of the conceptualisation may be demonstrated using the example of cities. Although cities enclose a large area, they may have to be represented using a point object in small scales. In this case, the abstractions of the spatial extent of the entity ends

[12] Cf. e.g. Peuquet (1984) for the spatial modelling of entities with vector data.

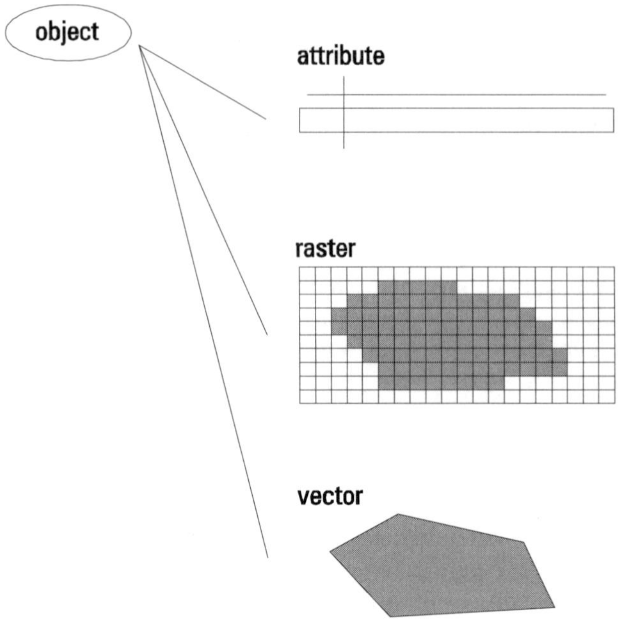

Fig. 2.14 Two spatial representations for a spatio-temporal GIS (adopted from Molenaar 1997:25 and 1998:32)

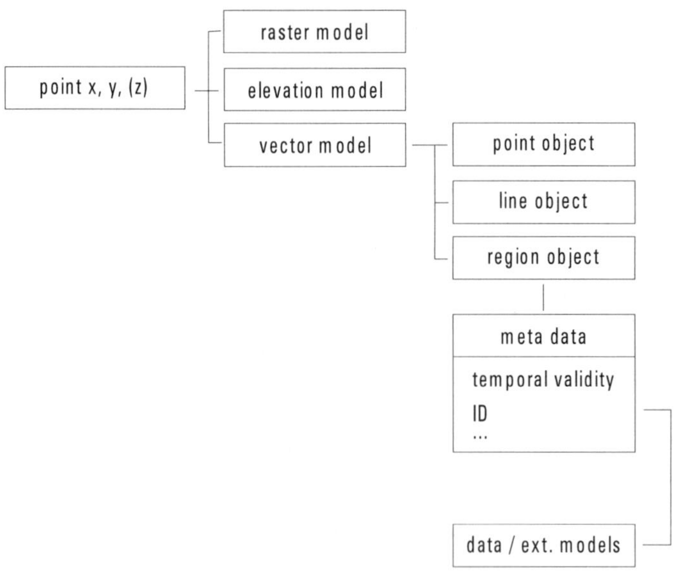

Fig. 2.15 GIS ER-model based on points (adopted from Molenaar 1997:45)

in a loss of information because the area of the city is lost as an inherent information when using a point object.

- *Line Objects* consist of at least two x-y-coordinate pairs. A complex line object may consist of more than one line segment. Each line object has a length and the spatial relation of line objects to other objects can be given in terms of left and right.
  Line objects are scale depended abstractions as well. As it applies to point objects there are no linear entities in the real world which do not have a physical extent. Line objects are commonly used for the representation of all entities which are of linear character, such as roads, railways, rivers etc. Again, areal information is not an inherent property of line objects.
- *Region objects* are defined by at least three pairs of x-y-coordinates. The coordinate points are organised in line segments which form a closed polygon. A region object has an area. The spatial relations can be defined as inside, outside, or adjacent.
  Areas modelled with a region object are the result of a discretization. There is no border visible in the physical world matching the border of a polygon. Regions are often used to separate administrative, statistical or electoral districts or units. Although there is sometimes a physical demarcation of such areal units existent in the real world, e.g. for cadastral purposes, in these cases the border is of abstract origin and may easily be defined by a polygon.

2.3.1.2
Logical data models for vector GIS

The organisation and storage of vector data is provided by the GIS software and cannot be changed by the user. Nevertheless it is important to understand how the data is structured, it affects the performance and the operations possible with the GIS. The logical data model may be studied from two perspectives. First, the way the geometries are stored in a database and second the method the GIS organises the objects and the linkage with attribute data (GIS architecture).

Each object is spatially represented by a geometry. A geometry is spatially defined by topology, shape and position (see figure 2.18). Topology gives information on the spatial relationships to other geometries, like inside/outside, left/right et cetera.[13] The shape defines the spatial extent and size of an area, the position defines the location of the geometry. The following three examples depict data structures applied in layer GIS:

- *Spaghetti structure:* Figure 2.19 shows, that for line and region objects only lines are stored. Each line consists of several points which are geographically defined by x-y-coordinates. In this case no topological structure is stored. If required, the spatial relationship between different objects has to be computed using geometric operations. As the shape of a region object is not given, in fact a region object cannot be differentiated from a line object.

[13] Cf. Zeiler 1999 on data structures.

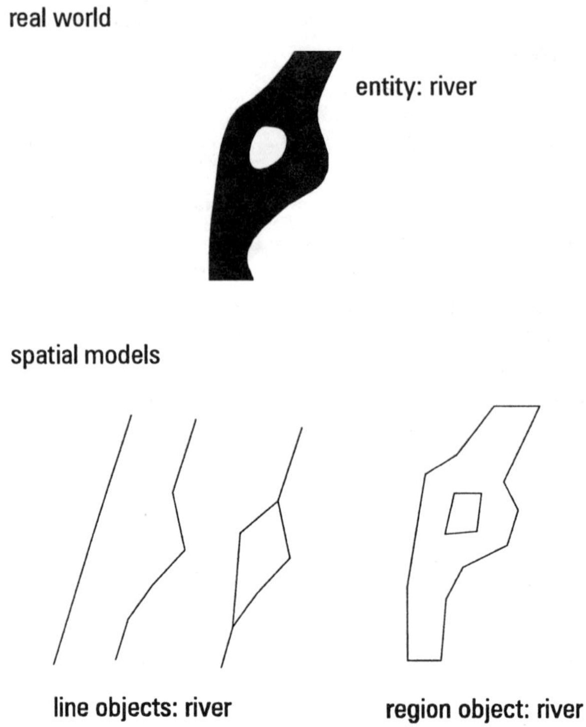

Fig. 2.16 Spatial models of a real world entity (adopted from Barteleme 1995:32)

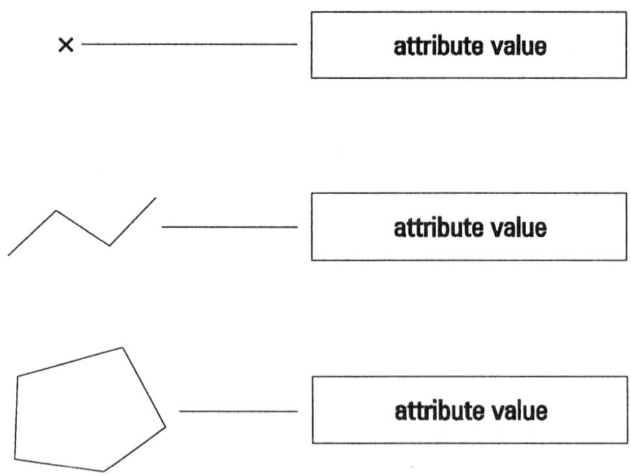

Fig. 2.17 Three geometrical object types in vector GIS (adopted from Molenaar 1997:25 and 1998:31)

- *Ring structure:* For the ring structure depicted in figure 2.20 a topological information is added. The stored data consists of single points as well. However, where two lines are intersected, an additional point is stored. Line objects and region objects are treated differently, region objects are used as closed polygons, having shape as a feature. Points that are part of different objects have to be stored redundantly.
- *Arc-Node structure:* With the arc-node structure topology is expanded in a way, that several spatial relationships are included in the data structure itself (see figure 2.21). Points are stored in a table defined by x,y-coordinates. Points indicating the beginning or end of a line or identical points belonging to different objects are called nodes. Nodes are stored in a tabular relation to a position given by a point in the point table. Line segments are called arcs; they consist of nodes and points stored in the tables. Only the relations are captured, no data is stored redundantly. Arcs are the basis for building line and region objects.

Using these structures, not only the way of logical structuring of geometry data but also how geometry data and attribute data are treated in a GIS can be differentiated. The differences are important for the functionality and performance of the GIS. The way the objects are organised by the GIS depends on the methods used to store spatial attributes of the geometries. Three approaches of GIS architectures can be differentiated (see figure 2.22):[14]

- *Layer oriented GIS:* In a layer GIS the logical structure of the geometric data is stored in a generic format. In most cases there is no interface available to access the spatial data directly. The attribute data is stored in a relational database and linked to the spatial objects using features identical in both data sources (IDs). The GIS functionality and the GIS front end are realised on separate levels.
- *Pseudo object-oriented GIS:* This approach stores the logical structure of the geometry data in a relational database. There is no difference between the geometric data and the related attribute data. The advantage of this approach is the open access to all data stored in the system. This method may be called a pseudo object-oriented GIS, following the nomenclature of object-oriented data structures. In a pseudo object-oriented GIS the logical data structures built in a relational database system is focused on single objects, not on layers. The GIS functionality and the GIS frontend are also realised on separate levels.[15]
- *Object-oriented GIS:* In an object-oriented GIS the data is not stored in a relational concept but in a real object environment where each vector object is linked to an appropriate class, which e.g. defines the functions (see figures 2.9 and 2.10).[16]

---

[14] Adopted from Molenaar (1997). He uses the terms "twofold", "stacked" and "integrated" with a slightly different meaning. Translation by the authors.

[15] An example of an object oriented approach with a proprietary DDMS is SMALLWORLD GIS Helokunnas (1994: 1198).

[16] Examples of this approach are e.g. GODOT (Günther & Riekert 1993), GeO2 (David et al. 1993), TIGRIS (Herring 1987), and OOGDM (Voigtmann 1997).

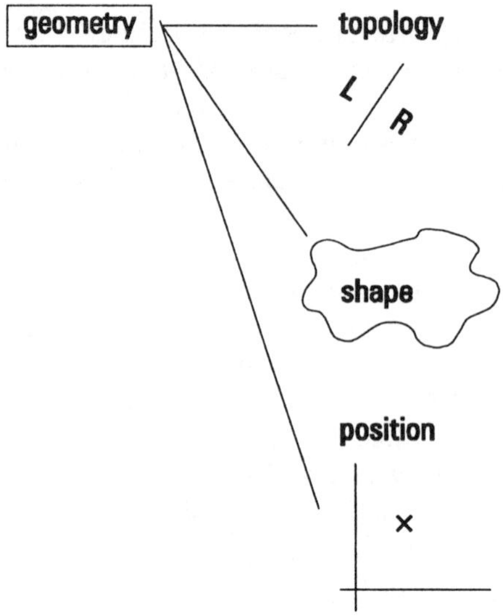

Fig. 2.18 Three aspects of geometry (adopted from Molenaar 1998:5)

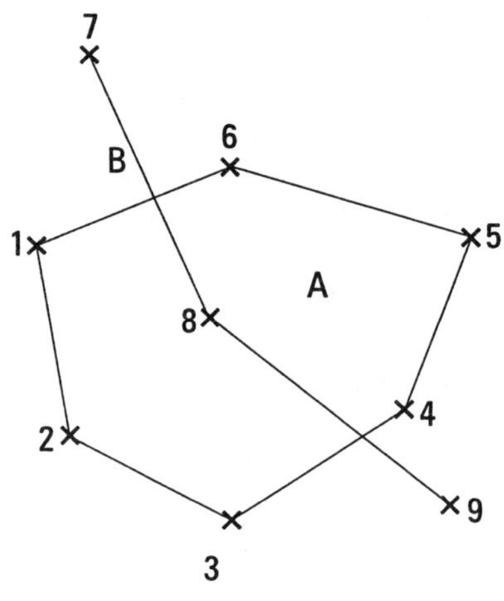

ID	Points
A	1, 2, 3, 4, 5, 6
B	7, 8, 9

Fig. 2.19 Vector GIS: spaghetti structure (adopted from Wessels 1997:52)

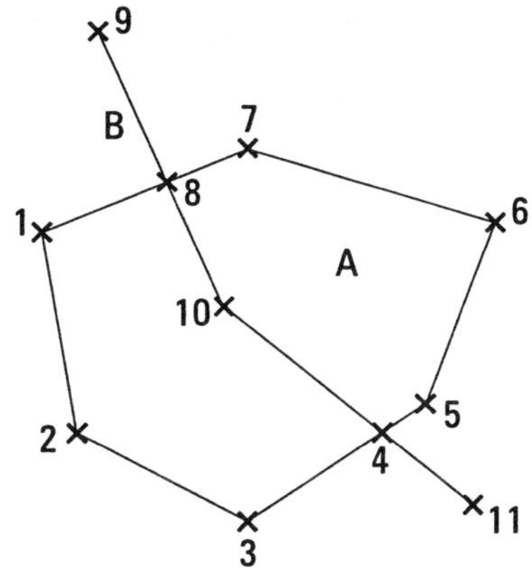

ID	Points
A	1, 2, 3, 4, 5, 6, 7, 8
B	9, 8, 10, 4

Fig. 2.20 Vector GIS: ring structure (adopted from Wessels 1997:52)

For the representation of spatio-temporal data the layer approach imposes several restrictions. The layer model focuses on the spatial representation of an object. Objects changing through time can only be stored in different layers. Without additional information stored in external database systems it is not possible to determine predecessor and successor. Furthermore, most layer GIS packages provide no ability to store information on the temporal validity directly with the geometry. In this case, temporal information can be represented by time slices or by an external database, storing the temporal validity for single geographical objects. This concept is linked to the least common geometry approach presented in chapter 4.4.6. As mentioned before, the majority of current desktop GIS use a layer concept.

The pseudo object GIS is in contrast able to include temporal relations between objects in a relational form. A spatio-temporal ER-model can be stored in a set of relational tables. This approach will be discussed in the following chapter. A practical approach is also based on this kind of GIS, which is discussed in chapter 4.4.7. Some professional relational database management systems provide extended features to deal with spatial data. In general, the GIS analysis and mapping functionality of a layer GIS may be used to simulate a pseudo object approach, if there is an open import-export interface. An external relational database may then be used to emulate a "geometry builder" as an input for the layer GIS.

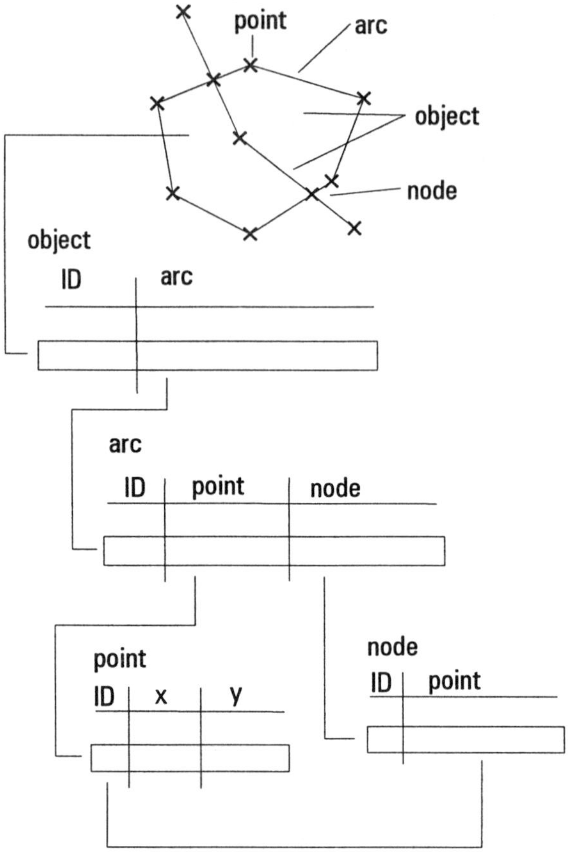

Fig. 2.21 Vector GIS: arc-node structure (adopted from Wessels 1997:55)

The system capabilities can be improved further by using an object oriented database or GIS to store the objects in a structure similar to the spatio-temporal ER-model. This makes the building and maintaining of the system easier and increases the system performance. Although object oriented database systems are available, they are not largely used for GIS yet. As object-oriented methods are predominantly concerned with programming concepts, we need to examine the potential benefits of these data structures for an average GIS user. Bartelme (1995:151-152) mentions the following differences of the object-oriented strategy as compared to traditional layer approaches:

- *Individualism:* Modelling is based on single objects, which are united with other similar objects in classes.
- *Renunciation of strict hierarchies:* Each object may be part of several classes.
- *Dynamics:* Instead of fixed layers flexible classifications are possible.
- *Behaviour:* Each object possesses rules of behaviour added to its attributes.
- *Areal coverage:* The layer approach implies that attribute values are available for every place inside the study area.

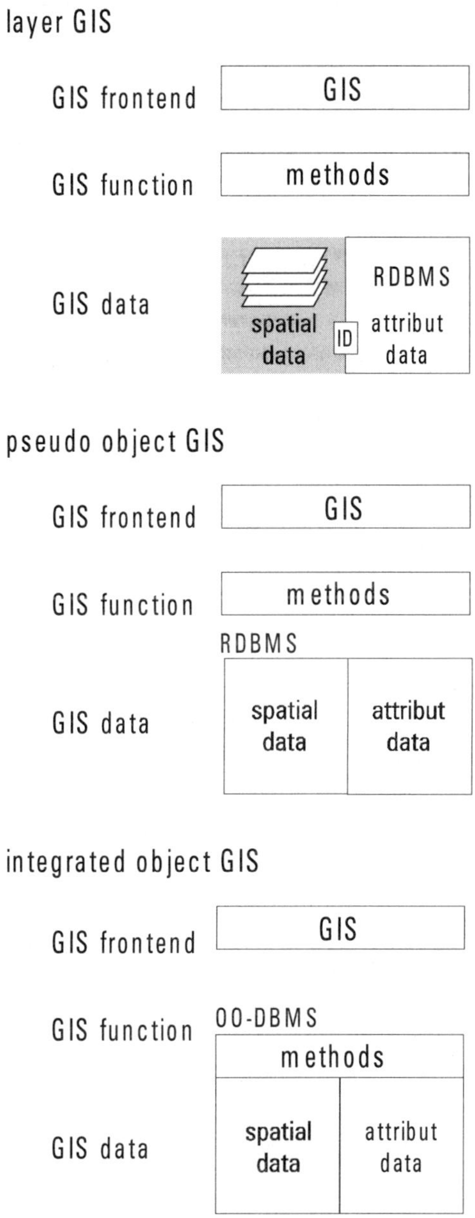

Fig. 2.22 Types of GIS architectures (adopted from Wessels 1997:62-63)

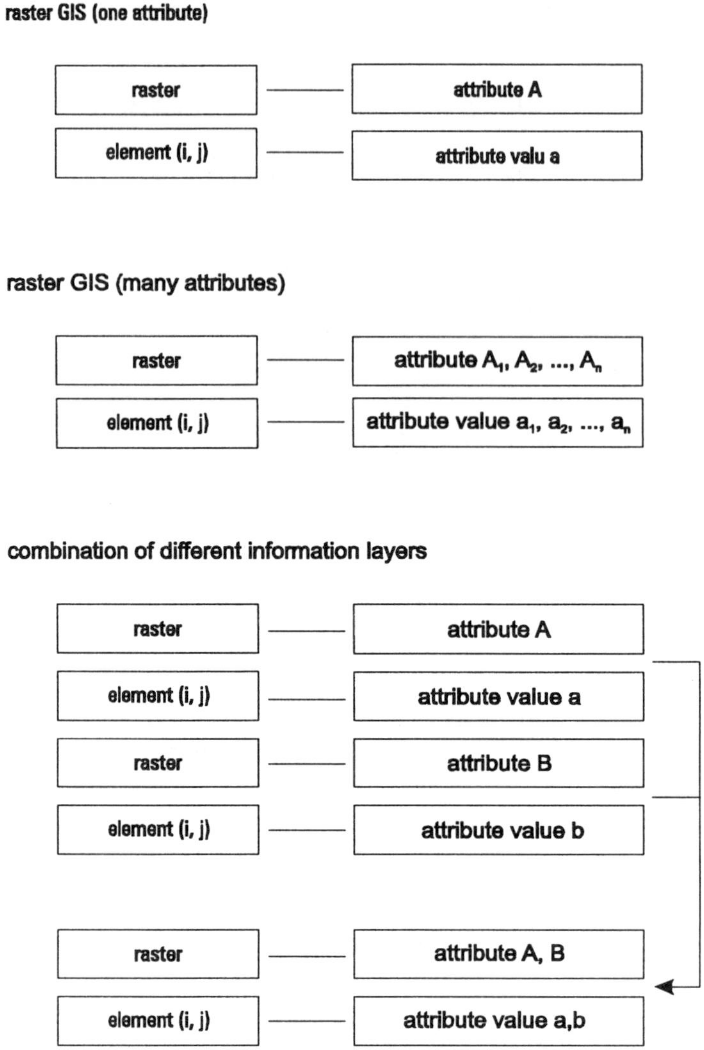

Fig. 2.23 Data structure of a raster GIS (adopted from Molenaar 1998:18)

2.3.2 Raster data models

Raster data structures use a direct position-attribute linkage to represent spatial phenomena. The continuum of the real world is transferred into finite and discrete elements using raster cells. Each cell is addressed as an element of the raster structure by a combination of row and column (see figure 2.23). Raster data may comprise single or multiple attribute values. Information organised in different layers of a raster GIS can easily be merged by combining the attribute values for respective grid cells. Raster with a different spatial resolution may be aggregated or dissolved to match the same resolution.

Raster data are often a result of scanning maps. They may be produced directly with digital cameras or sensors, e.g. in form of satellite images. More sophisticated logical raster data structures aim at the reduction of the great amount of storage space needed as a result of the redundancy. At this point, raster data structures cannot be discussed in great detail.[17]

The ability to represent spatio-temporal entities is restricted by the coarse conceptualising of the spatial representation of linear and areal objects and the lack of topological content. In fact it is not possible to determine whether two adjacent raster cells belong to the same linear or areal entity. Measures such as length or area which can be calculated for line and region objects in a vector environment are not possible in raster GIS. Similar to the spatio-temporal representation of the layer concept, raster GIS are focused on the spatial aspect. The position of the raster cell is treated as invariable, attribute data and temporal information are assigned to the raster cell. Temporal information as well as different levels of content may only be assigned using raster layers. Nevertheless the raster data format can be of great advantage for the representation of spatio-temporal data, as they provide an easy way to detect changes between two layers (see chapter 5.3).

2.4 Integration of external models in Geographic Information Systems

As discussed above, external models are used for the conceptualisation of the real world and the definition of entities. When such entities are modelled and transformed into objects that can be stored in a GIS, they have to be evaluated in terms of their influence on the analysis which shall be applied in later steps of the implementation process. External models have to deal with the relationship between observed attribute data and the respective location. The problems of observing features can not be covered in this book. The reader has to refer to readers in methodology and techniques of their respective field for this purpose (see chapters 1.3 and 9.1).

For the conceptualisation and implementation of a GIS it is often not possible to acquire data on observations of individual entities. The initial observations are often provided in a pre-processed form, as aggregated data, e.g. summarized or averaged individual observations for an administrative unit or other areas. Statistical data is always aggregated data and based on the artificial subdivision of space. As most GIS applications depend on the use of such sets of attribute data and the related territorial area units, the external models used to produce such data must be assessed.

---

[17] Cf. e.g. Peuquet (1984) on raster data structures.

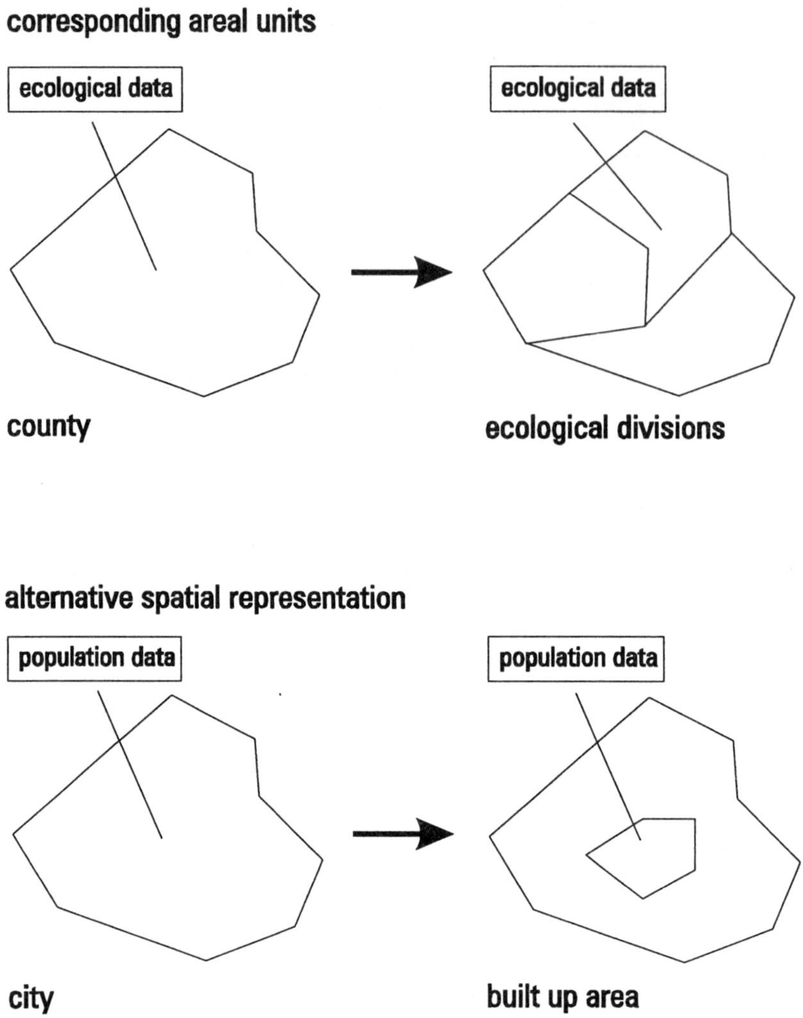

Fig. 2.24 Spatial representations of attribute data

There are two kinds of effects which must carefully be observed in terms of the influence on the building of objects and establishment of relations with attribute data. One effect evolves from the fact, that the meaning, content and interpretation of individual data aggregated for areas is affected by the selection of the areas. Two examples are given for this kind of effect. The dasymetric mapping problem and the modifiable areal unit problem. The other kind of effect is caused by a misconception of measures used to represent individual observations in aggregated form. Also two examples are given, the ecological fallacy problem and the closed number set problem.

2.4.1 The dasymetric mapping problem

In cartography percentages or densities are often represented using choropleth maps, where the areal units are shaded according to the classified data to be mapped. Obviously the advantage of this method is, that it is easy applicable to all kinds of statistical data like survey or census data using given territorial units such as census tracts or municipalities. In fact this technique is often misleading as already shown by Wright in 1936 (Wright 1936). In his example he mapped the population density of Cape Cod, using entire townships, townships without the uninhabited areas and recalculated figures for population density, and subdivisions of townships as mapping units. The resulting maps show that choosing the appropriate mapping units associated with the data to be mapped is important for the interpretation of the inherent spatial structures of the data (Jenks 1963).

Using desktop mapping or Geographic Information Systems the easy availability of such standard areal units as census tracts, postcode areas et cetera seduce to an uncritical usage of inappropriate spatial divisions of the study area.[18] For the modelling of spatio-temporal processes it is necessary to use spatial objects with a territorial division which corresponds to the content of the data (see figure 2.24). A second mistake is encouraged by the easy availability of standard administrative units for GIS. The use of such standard areal units may lead to misinterpretation when they are inhomogeneous in terms of the data to be represented by the units. Ecological data should e.g. not be mapped using administrative boundaries including different types of landscape units. In some cases it is necessary to use formally invalid areal representations in order to meet this criteria. Population data for an administrative unit may for example sometimes better be applied to the built up area to calculate population density measures. This question is highly scale dependent. Areal units which are too large and therefore suffer from inhomogeneous distribution of the attribute values assigned to them, are often used because of their availability. The same applies to the problem of aggregation of data so, that the attribute data fits existing geometry data sets.

[18] Cf. Plane & Rogerson (1994:353-354) on how to deal with the dasymetric mapping problem for population density maps.

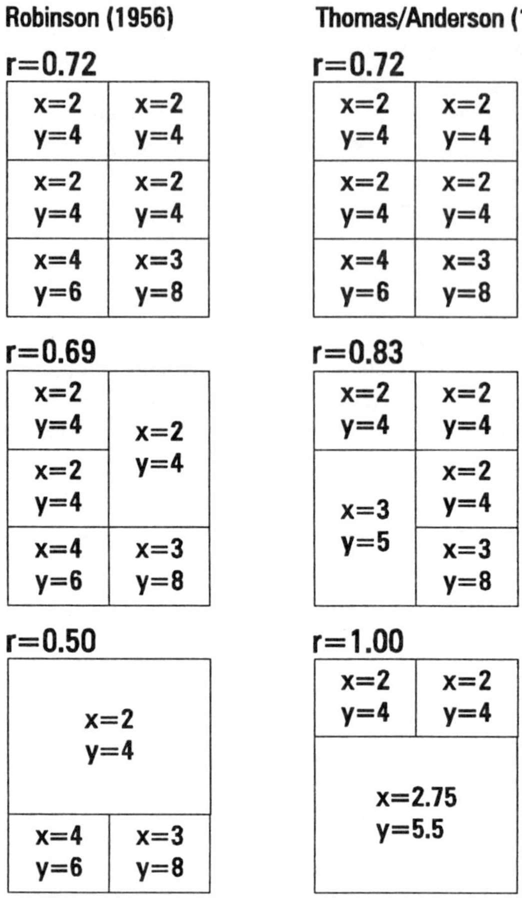

Fig. 2.25 The modifiable areal unit problem (Robinson 1998:89)

2.4.2 The modifiable areal unit problem

The modifiable areal unit problem is widely known in quantitative geographical research, but in fact often enough not properly acknowledged during the design of a research project.[19] The key point of this problem is that geography matters. In his research on the correlation of annual rainfall and the density of farms per square mile in the counties of Nebraska, Robinson (1956) showed that the size and shape of the areal units under study affect the outcome of the correlation coefficient analysis between the variables computed for a number of counties.[20]

[19] Cf. Arbia (1989), Kollarits et al. (1998) and Openshaw & Alvanides (1999) on the modifiable areal unit problem. See Martin (1998) on the building of areal units for statistical purposes.

[20] The example is taken from Robinson (1998:89f).

As shown in fig. 2.25 an even distribution of the size of the areal units results in a relatively high correlation coefficient of 0.72 (example 1). The aggregation of two of the counties decreases the correlation between the two variables to 0.69, in a further stage the aggregation of four counties to one areal unit and only three remaining units results in a relatively low correlation of only 0.5. Robinson assumed that this size dependent result of the correlation coefficient may be dealt with by using an area weighted method for the computation of correlation coefficients. In fact not only size and number of the areal units have a crucial impact on the calculation of the resulting correlation. Thomas and Anderson (1965) prooved, that an alternative aggregation of the original counties used by Robinson leads to a completely reversed effect on the resulting correlation coefficient. As seen in example 2 the numbers for the correlation increased to 0.83 and 1. A conclusion which had to be drawn from this result is, that the shape of boundaries used to define areal units in a correlation analysis will affect the outcome in an almost unpredictable way. By defining areal units in which way ever, one includes the effects of other underlying variables which are co-measured in the correlation analysis. In this case undeclared assumptions on spatial structures under research have been "imposed on the system in order to measure it", as Kendall & Yule (1950:312) stated. In fact the practical consequences of this problem have long been known by many electoral geographers. By carefully choosing an appropriate zoning system of electoral districts the outcome of elections could easily be manipulated, known as the gerrymandering problem.[21]

The knowledge of this effect is of vital importance for the construction of all areal unit systems used in geographical research and models of real world entities in Geographical Information Systems. The effects of the modifiable areal unit problem may be subdivided into three main aspects (Openshaw & Talyor 1981, Openshaw 1984, Robinson 1996):

1. The difference between modifiable and un-modifiable areal units relates to the problem of areal definition, used for statistical purposes. Some areal units are defined, like administrative boundaries, others are chosen by researchers in order to meet their requirements. In both cases the ecological fallacy problem applies (see chapter 2.4.4).
2. The scale effect of the modifiable areal unit problem implies, that the larger the units used for a aggregation procedure are, the higher the correlation between the two variables is on average.[22] The common application of aggregation of different levels of scale have to take this fact into consideration.
3. The zoning effect of the modifiable areal unit problem. A given area can be divided into an infinite number of alternative spatial divisions (see figure 2.26). The areal units may be given a priori, such as administrative units, census tracts etc. or some kind of rules could be applied to establish a grouping or zoning system. The outcome of a correlation analysis depends on these options.

---

[21] Cf. Openshaw & Taylor (1989:61) for a brief survey on the gerrymander problem and different zoning approaches for electoral purposes.

[22] Cf. e.g. Openshaw (1979) and Woods (1976) on the scale problems.

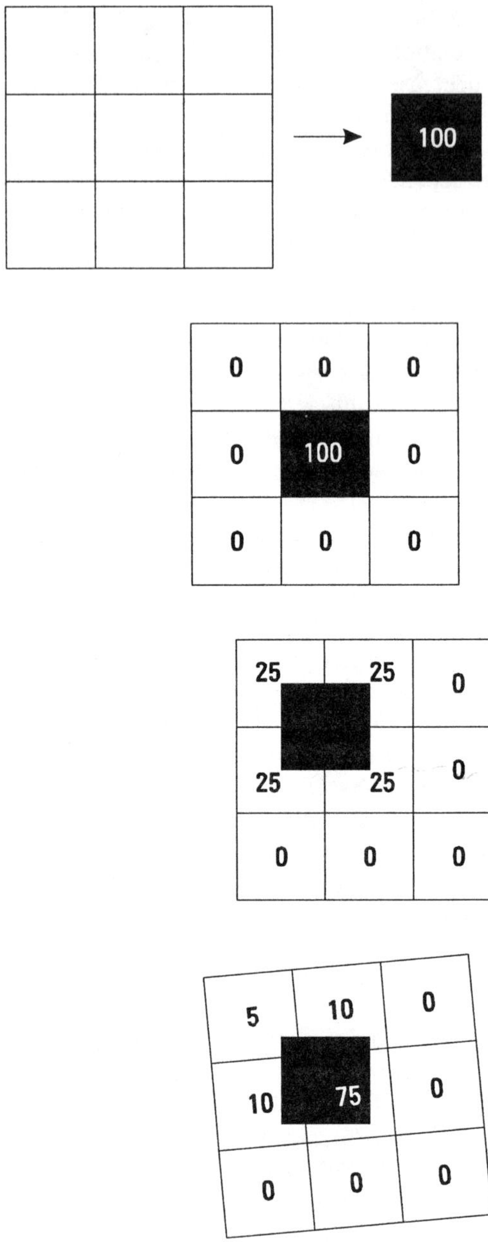

Fig. 2.26 Zoning effect of the modifiable areal unit problem

There are several ways to deal with the problem, described in the statistical textbooks (e.g. Openshaw 1984). However, at this point the different approaches can not be discussed in detail. For the building of a spatio-temporal GIS the modifiable areal unit problems have to be considered, especially if correlation analysis shall be used as an analysis technique. This is of great importance, because the digitisation of areal units for a GIS is time consuming and expensive; a later adoption to another, more appropriate design of entities and their spatial representation is hardly possible.

2.4.3 The closed number set problem

Closed number sets are measures like percentages or proportions. Dealing with spatio-temporal data a transformation in such closed number sets is often applied, so that data for uneven areal units can be compared directly. If such data is used for further statistical calculations, the result of measures like the correlation coefficient may be affected unpredictably (see figure 2.27).[23]

an open number system

Region	Land Type (100km²)			
	Arable	Pasture	Scrub	
	X	Y	Z	r_{xy}
A	5	2	1	
B	7	4	2	0.99
C	8	6	4	0.96
D	10	8	5	0.89

a closed number system

Region	Land Type (%)			
	Arable	Pasture	Scrub	
	X	Y	Z	r_{xy}
A	63	25	12	
B	54	31	15	-0.97
C	44	33	23	-0.98
D				0.89

Fig. 2.27: Example of the closed number set problem (Robinson 1998: 91)

Calculation measures like percentages or relations will always lead to a loss of information. A second information is always needed to recalculate the original data value. Therefore it is a common advice, not to use closed number set data for building a spatio-temporal data base if possible. It is preferable not to use such

[23] Cf. Robinson (1998: 92) for an example how to deal with this problem in statistics.

closed number set data for further statistical processing. If measures like percentages, densities et cetera are necessary to interpret the data, they can be derived in the GIS for the purpose of mapping only.

2.4.4 The ecological fallacy problem

The availability of aggregate data for areal units e.g. from census statistics or the need to provide confidentiality has fostered a widespread use of aggregation of individual data in geographical research. The aggregation of individual data is in fact sensitive against the distortions produced by the structure of the spatial division of the research area used for aggregation. [24] Robinson (1950) correlated the share of illiteracy and foreign origin for the nine U.S. census divisions resulting in a correlation coefficient of –0.69. The ecological fallacy, occurring when aggregated data is used to draw insight on characteristics of individuals, results in a positive correlation of +0.118 in the same case when the analysis is performed on the basis of individual data.

Using aggregated data the spatial distribution of the variables among the areal units affects the computed correlation coefficient. As a result, one can conclude that it is not possible to gain insight on the relationship between variables on the individual basis by analysing data which is aggregated to areal units (Robinson 1998: 91).

The problem addressed by the ecological fallacy problem should be considered for the collection, aggregation and interpretation of spatio-temporal data. Following the research on the ecological fallacy problem other fallacies were detected as well (Alker cited after Johnston 1996:145):

- *Individualistic fallacy*, which means that the whole could be more than the sum of individuals.
- *Cross-level fallacy*, which addresses the transferability of observations between different levels of scale.
- *Universal fallacy*, which implies that observations made in a given sample may not by valid for the whole.
- *Selective fallacy*, which states that selected cases canot justify a hypothesis.
- *Cross-sectional fallacy*, which addresses the transferability of observations between different time steps.

[24] Cf. Arabia (1989:21) on the ecological fallacy problem.

2.5 Spatio-temporal dimensions in GIS

The kind of data which can be represented in a GIS may be classified in dimensions (see figure 1.2).[25] In GIS terminology four dimensions are differentiated. At least one additional dimension must be added for the content which is used in the GIS. The concept of dimension is essential for the use of GIS and their design and implementation. The organisation of the data in databases and GIS depends on the structuring of the features of entities in dimensions. For example, a table is only able to address three dimensions in the form of rows, columns and data fields. For the analysis and the display of the remaining dimensions the information must be structured otherwise. A practical approach to deal with this problem is presented in chapter 5.

- *Two dimensional (2-D) GIS*
  The first two dimensions of the data matrix in a GIS cannot be addressed separately. A flat surface is described by the two coordinates, x and y, measured e.g. by longitude and latitude. A third dimension can be simulated by using x-y coordinates, e.g. as contour lines in topographic maps.
  For the purpose of presentation only two dimensional data can be mapped directly. All other dimensions have to be symbolized, so that they can be mapped on the two dimensional surface of a paper map or a computer screen.
- *Three dimensional (3-D) GIS*
  A third dimensions is added with an additional z-value fur each x-y coordinate, in order to model elevation. In fact it is not possible to visualise a real volumetric 3-D surface directly. All attempts have to deal with the limitation of a two dimensional presentation.
  Perspective views of 3-D data sets are called 2,5-D presentations. Real 3-D GIS treat the third dimension as volumetric objects. Such volumetric models are more versatile and are suited much better for analysis and visualisation, but nevertheless they must be reduced to two dimensions for visualisation.
- *Four dimensional (4-D) GIS*
  The fourth dimension is time. Every entity in the real world can only exist, if it exists in time. Besides the three spatial dimensions another temporal dimension is always existent, but often neglected. Time can also be modelled so that it can be stored in a database. This is discussed in detail in chapter 3.
  As well as the third, the fourth dimension cannot be visualised directly in a GIS. Using the computer a four dimensional animation may display spatio-temporal processes. In two dimensional maps not only the third but also the fourth dimension must be mapped using a modelled symbolisation of time or temporal changes.
- *Five dimensional (5-D) GIS*
  The fifth dimension is the attribute data value, which is used in most GIS to carry meaning or as the subject of analysis and display.

[25] Cf. Davis & Williams 1989

Concluding this, the dimensions in a GIS can be defined by x, y ,z ,T and A-values (see figures 2.28 and 2.29). In a GIS not all dimensions must be applied. In some cases of spatio-temporal GIS the third dimension is not used. To call such a GIS a 4-D GIS could be misleading but in fact this term is commonly used for temporal GIS.

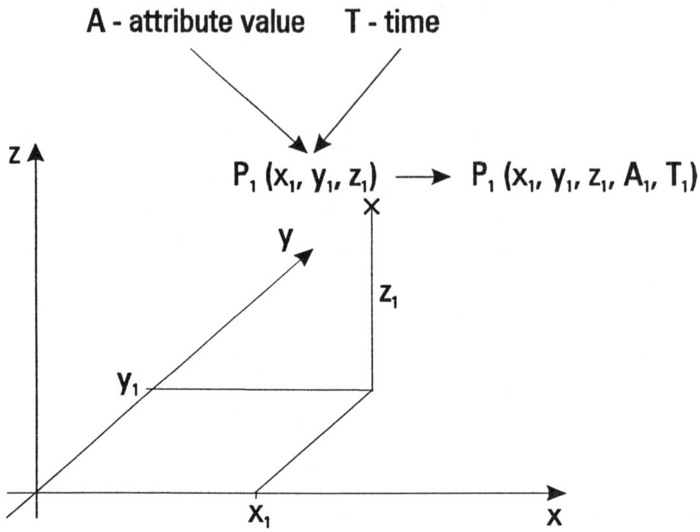

Fig. 2.28 Point in five dimensions (adopted from Davis & Williams 1989)

dimension	values		type
second	x-coordinate	longitude	spatial (area)
	y-coordinate	latitude	spatial (area)
third	z-coordinate	elevation	spatial (volumetric)
forth	T-value	temporal indicator	temporal
fifth	A-value	attribute data value	content

Fig. 2.29 Dimensions in GIS (own design adopted from Davis/Williams 1989)

3 Integrating time in Geographic Information Systems

> "Time may basically be thought of as temporal relationships between events and objects in space. Time and space pervade all aspects of reality that constitute the data of geography, since everything material must exist in time as well as 3-dimensional space."
>
> (John Cole and Cuchlaine King 1969:405)

A spatio-temporal GIS extends a customary two or three dimensional spatial GIS on the time axis. Hägerstrand (1970) already used an orthogonal relationship between space and time to conceptualise the framework of his Time Geography approach. For his analysis of space-time paths, he made use of the fact that people can only be at one location at a time and that the moving from one location to another needs time.[26] A space-time path is defined by the changes of locations over time (see figure 3.1). This example shows, that in a spatio-temporal framework space and time are equivalent features of an event or a process. Furthermore each entity which physically exists in the real world is defined in a spatio-temporal context which can be conceptualised using spatial, temporal and attribute information (see figure 1.2).

"The concept of time is intuitive. We use it in many ways in our daily live. The passage of time is normally understood via changes we perceive occurring to objects in space – their transformation over time and their movements in relation to one another" (Peuquet 1994:441). In the previous chapter the conceptualising of the spatial dimensions of entities was discussed. It was stated, that entities must be spatially modelled in order to be used in a GIS. The same applies to a time-integrative GIS. The increasing levels of abstraction used as a guideline for the building of a spatial representation of the real world applies to the temporal aspects in a similar way. In complete analogy to space, time can be understood as an a priori continuous and infinite dimension, which must be reduced in complexity so that it can be stored in a GIS as finite and discrete data. This aspect is discussed in chapter 3.1.

Every spatial object used in a GIS has a temporal validity as well as one or many attribute values. The entity of a spatio-temporal process may change its spatial representation over time as well as its spatial relationship to other entities. In addition, the related attribute information may be subject to changes throughout

---

[26] Cf. Wachowicz (1999:19) on the analysis of space-time paths with GIS.

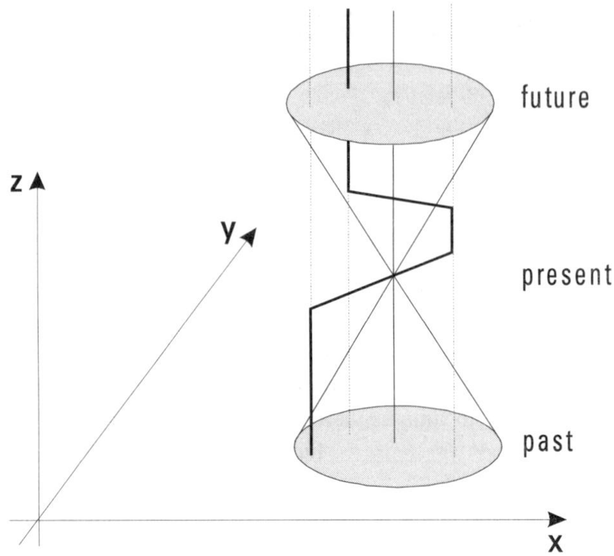

Fig. 3.1 Space time path (adopted from Peuquet 1994:453)

time. All spatial objects in a GIS are in the first instance defined by their spatial representation (see figure 2.25). This concentration on the spatial aspect of an object as the focal point of the conceptualisation of a spatio-temporal process in a GIS neglects the fact, that time and space are equivalent dimensions. The questions of what, where, and when defined as the basic components of the triadic framework are not interchangeable and equally important for the understanding of spatio-temporal structures and processes (see figure 1.1). Section 3.2 gives an overview of different approaches of the conceptualising of time related to spatial objects in a GIS. In this context "the concept of time implies that changes occur throughout the present, the past and future of the life span of a real-world phenomenon. A temporal GIS will aim to understand these changes and their effects over time rather than simply reproducing them by displaying a sequence of snapshots" (Wachowicz & Healey 1994:106).

According to Langran (1993:5) the following major functions of a temporal GIS can be distinguished:

- *Inventory:* Storing a complete description of the study area, and accounting for changes in both the physical world and computer storage.
- *Analysis:* Explaining, exploiting, or forecasting the components contained by and the processes at work in a region.
- *Updates:* Superseding outdated information with current information.
- *Quality control:* Evaluating whether new data are logically consistent with previous versions and states.
- *Scheduling:* Identifying or anticipating threshold database states, which trigger predefined system responses.

- *Display:* Generating a static or dynamic map, or a tabular summary of temporal processes at work in region.

To realise these functions in a conventional GIS, temporal information for objects must be added to the logical data models used in GIS. Different approaches to do so are discussed in chapter 3.5. "Thus, while great progress has been made in developing data models for GIS that go beyond the time slice approach, the creation of a truly spatio-temporal GIS remains an unmet challenge" (Couclelis 1999:35). The integration of time in existing desktop GIS packages is an even greater challenge today. In chapter 4 the treatment of this aspect will be succeeded by the introduction of practical approaches which can be applied using conventional technology at hand at the moment. As a prerequisite for this attempt basic concepts like event time and database time as well as different methods of versioning of databases and GIS have to be utilized. These basic concepts are explained in sections 3.3 and 3.4.

3.1 Different types of time

Time as well as space is a concept we instinctively use to understand and conceptualise our environment. Because we do not posses a sense to perceive time directly, we need to conceptualise the nature of time: Time structures several aspects of our perception. Time describes when something happens. In this sense time describes a certain event in a continuous dimension. The occurring of events could be measured in different ways. However, we experience not only that something happens but also that time passes by and that there is a duration between two incidents. A period of time can also be measured in different ways. From our understanding of the real world we may derive the concept, that time is closely related to changes, because every change, in what ever way, can only take place if time passes by. The complex nature of time may be added to a GIS by using additional concepts or perspectives. Although such aspects may lead to interesting insights, the options for the integration of temporal information in a GIS are restricted. Hence, time may be included in several ways (Laurini & Thompson 1992:104):

- as a basis for recording events or attributes,
- as an attribute of an entity with unchanging spatial properties, or
- as a framework for observing changes in the spatial entities.

According to the taxonomic model of Frank (1998; see figure 3.2) time can be conceptualised either as a linear or a cyclical dimension.[27] A linear treatment of time implies a consecutive development on the time axis, while a cyclic treatment conceptualises the fact, that time may have a reoccurring character. The measure of time can also be differentiated according to whether time is transferred to a discrete scale or treated as an undivided continuum.

[27] Cf. also Charalampos & Loucopoulos (1991)

<table>
<tr><th colspan="2"></th><th>Total order</th><th>Partial order</th><th>Branching</th><th>Multiple</th></tr>
<tr><td rowspan="2">Linear</td><td>Ordinal</td><td>Single experience</td><td>Multiple experience</td><td rowspan="4">Branching time</td><td rowspan="4">Time with multiple perspectives</td></tr>
<tr><td>Continuous</td><td colspan="2">Continuous time</td></tr>
<tr><td rowspan="2">Cyclic</td><td>Ordinal</td><td colspan="2" rowspan="2">Cyclic time</td></tr>
<tr><td>Continuous</td></tr>
</table>

Figure 3.2 Taxonomic model of time (Frank 1998:45)

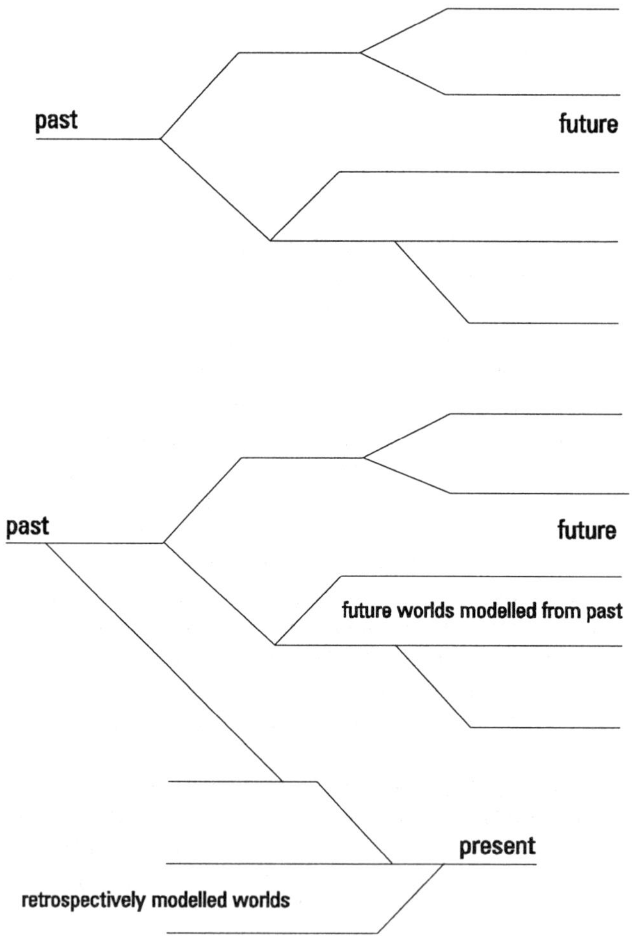

Fig. 3.3 Bi-directional branching time structure (adopted from Hazelton 1992:305)

Events measured on a linear ordinal scale can be differentiated further, according to the relationships of events. In a *total order set*, all relations between all events are explicit. In a *partial order set*, the relationship between some events may be given, the relationships between others are not known. On the other hand it is also possible, that there is more than one time line leading from present to past and future (branching time). This is a result of the fact that an event may occur more than once. The simulation of different future states is a potential example (see figure 3.3). On the other hand it is also possible to derive retrospective models from the present state (see chapter 4.4). From such a past state also different alternative future states may be simulated. In figure 3.1 the potential spatial range of events and processes for past and future time slices are depicted, evolving from „here and now" (Peuquet 1994: 453). Time with multiple perspectives refers to the fact, that „temporal models of reality cannot totally be synchronised with reality" (Frank 1998:54) (see sections 3.3 and 3.4).

For practical reasons it is necessary to use discrete measures in GIS. Calendar time is not the only kind of time possible for this purpose (Frank 1998:41). A lot of temporal information may not be addressed by a certain date, e.g. in geology or archaeology. The measurement can be made on different scale levels, reaching from an ordinal to an interval scale level. On the interval scale level of time discrete intervals with equal duration are used. The shortest interval used as a measure for a period and for the definition of the scale intervals used to define the time of an event on the time axis is called a chronon. The granularity of time, i.e. the smallest chronon applied, defines the level of precision and uncertainty with which time can be depicted as shown in figure 3.4.

The length of a time span depends on the start and end, given by events. The length is also dependent on the granularity which is used to measure time. If for instance hours instead of days are used as the smallest chronon, the spatial resolution increases and the level of uncertainty decreases (Frank 1998:43; Kemp & Kowalczyk 1994:91). As time may always be divided further, no smallest chronon can be defined.

In practice the determination of an appropriate granularity depends on the scale of the time axis, which is defined by the duration of periods to be measured. As with spatial information, for temporal information scale dependency matters as well. Geological information can e.g. be measured with 1,000 years as the smallest measure, while data captured for meteorology must be measured using hours or minutes as a chronon.

When cyclic time is used, it is not possible to calculate time against an origin. Nevertheless it is possible to arrange the events according to a time scale with predefined equal intervals. If this is the case, e.g. using the hours of a day or the days of a week etc. a time span between two incidents may be calculated from cyclic time similar to linear time.

Using time scales with different chronons, it is possible to aggregate time measures, e.g. from hours to days or from months to years. The other way round, a temporal decomposition, is also feasible. In this case a suitable method of disaggregation must be used (see figure 3.5).

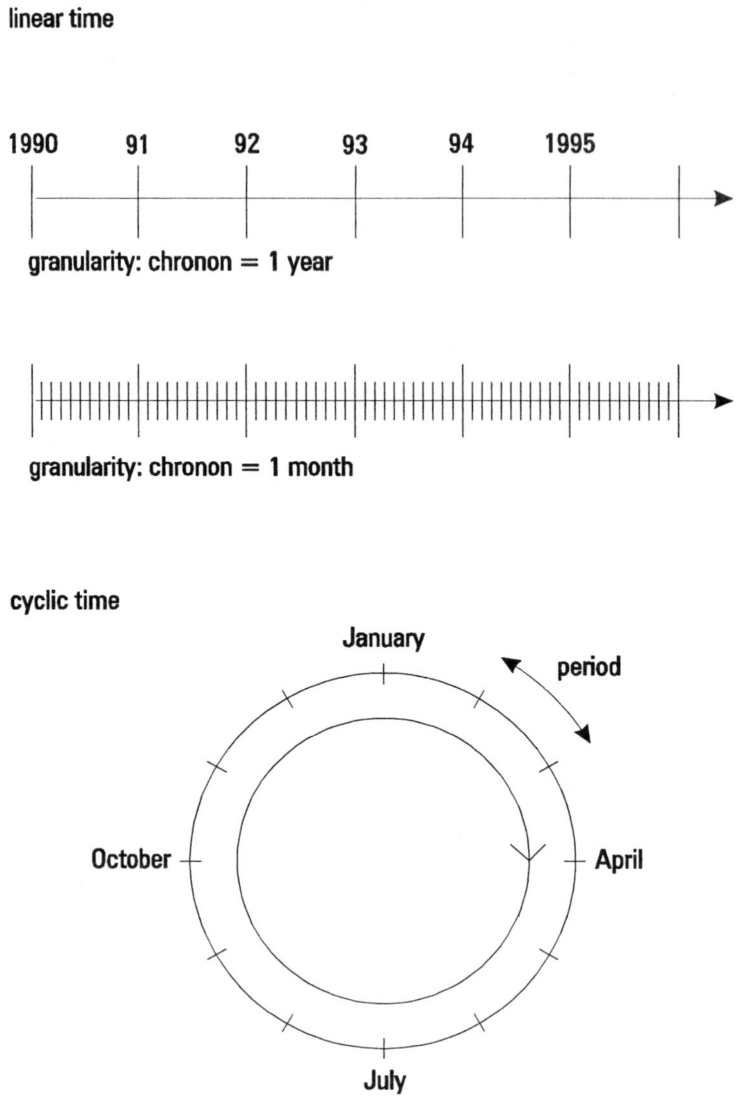

Fig. 3.4 Granularity of linear and cyclic time (adopted from Frank 1998:52-53)

Finally, the question of time discretization is of great importance for the integration of time in a GIS. For most purposes it is too expensive or not possible to treat temporal phenomena in a continuous way. Depending on the size of the chronon used to define the time scale, measurements or readings from a data logger cannot be taken incessantly. In such cases it is important to chose a time sequence between two measures which corresponds both, with the time scale and the nature of the events or processes to be dealt with (Laurini & Thompson 1992:105).

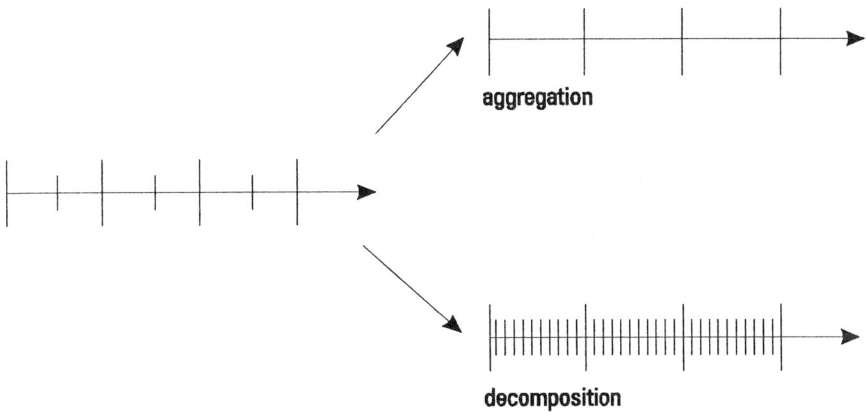

Fig. 3.5 Temporal aggregation (adopted from Lee & Kemp 1998:156)

3.2 Conceptualising time in GIS

Time can be conceptualised in different perspectives. While some conceptualisations could rely on non temporal spatial objects, for most views on the world objects will also change their spatial representations trough time. Spatio-temporal changes may occur in different types and forms (adopted from Snodgrass 1992:35):

- Attribute and spatial object is invariant, the GIS spatially represents only one point in time or a past in present view based on one time step.
- Attribute changes, spatial object is invariant, the GIS references different temporal steps of attribute data to only one spatial state represented by a time slice,
- Attribute is invariant, spatial object changes, the GIS processes spatially changing objects but only temporal invariant attribute data is available.
- Both, attribute and spatial object change, the GIS needs to join changes of spatial objects and attribute data which could proceed independently from each other.

In the following sections approaches and concepts to conceptualise time are explained. Using a relational database three alternative methods to build the conceptualisations in a GIS may be distinguished. In general, temporal information has been incorporated into GIS spatial data models by time-stamping single layers, attributes and spatial objects (Yuan 1999:2).

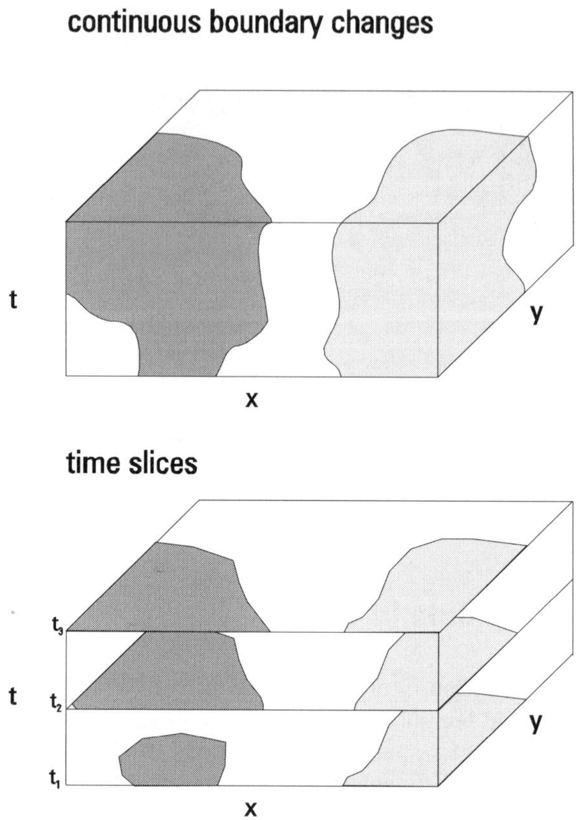

Fig. 3.6 Space-time cube (adopted from Langran 1993:30)

3.2.1 Snapshot

The snapshot of a real world phenomenon is the most simple case of a temporal GIS. Using a mono-temporal set of layers is the standard case in a two dimensional GIS. Temporal content may be represented by assigning temporal data to the layer. The nature of temporal data used is not restricted to mono temporal data. Data representing temporal measures for elements of the selected time slice (past in present) may also be integrated. The temporal representation of data must match the time valid for the time slice, or lie within the validity of spatial objects represented by the layer. This basic concept can be implemented easily and will be used for mapping temporal information in chapter 6.

3.2.2
Time slices

Continuously changing borders, e.g. as a result of a process, may be transformed in discrete sections using the time slice method (see figure 3.6). The time steps for which the layers are created may be chosen either periodically, at the occasion of specific events, or randomly. Thus, time in a layer oriented GIS environment is represented by a set of layers in which each layer contains spatial objects of the same category valid at a given time step (see figure 3.7). This concept depicts the distribution of temporally „homogeneous units" in one layer with no temporal relations between the layers (Yuan 1999:2). Using this approach we are not able to gain evidence of events or changes that have happened in-between two time-stamped layers. This method of temporal map sets (Beller et al. 1991) stores the spatio-temporal data in a redundant way.
Temporal information may be integrated matching the changing spatial states of successive temporal layers, or a spatially invariant layer may be used to display different time slices of changing attribute data. Time slices may also be used to perform geometrical operations, in order to derive changes between the time steps represented by a set of layers (see figure 3.8).

The time slice method is frequently used for raster GIS, where the aforementioned geometrical operations to detect changes are easy to apply (see chapter 5).

3.2.3
Topology of time

The topology of cartographic time as it was introduced as a concept by Langran (1993:32-33) prooves the fact, that layers may be used to produce a quasi continuum of spatio-temporal changes of spatial objects (see figure 3.9). Langran uses objects with versions valid between two incidents of mutation to conceptualise the time axis. If several objects with their different versions and time of mutation are overlaid and intersected, a map results with valid states and events of change. If each possible mutation is recorded and the respective objects are intersected with each other the resulting layers will represent a continuous record of changes simulated by time slices.

This approach can be applied if changes are of discrete nature. Changes of administrative boundaries et cetera which are often used for socio-economic applications, may be treated in this way. Although it is possible to reconstruct all occurring changes with a set of time sliced layers, a more sophisticated design should regularly be used in such a situation. Without much additional effort, the strategies described in the following sections provide better ways of analysis (see chapter 5).

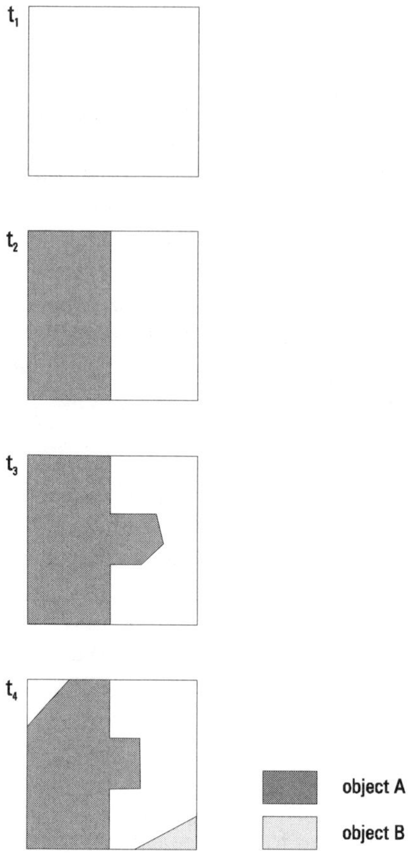

Fig. 3.7 Snap shot time slices
(adopted from Langran 1993:39)

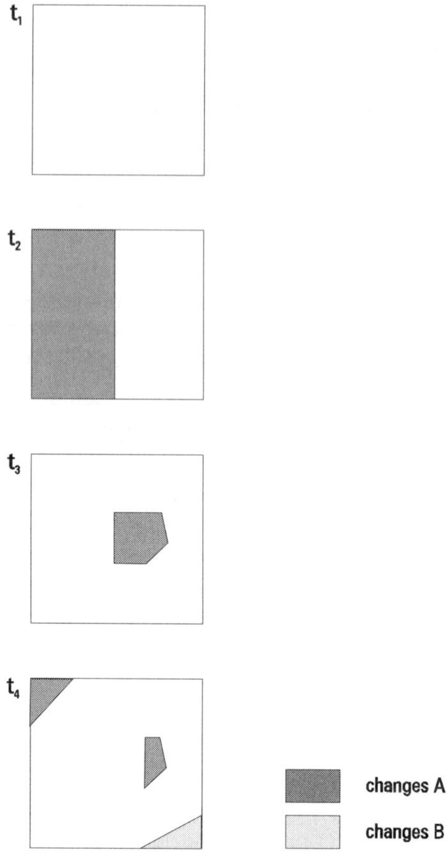

Fig. 3.8 Temporal changes of times slices
(adopted from Langran 1993:40)

3.2.4
Space-time composite

Figure 3.6 shows how borders contained in a space-time cube can be transferred into a set of overlaying layers. If the concept of spatial topology is applied on the way in which the layers are used for the discretisation of the continuous space-time cube, a space time composite with „spatially homogenous and temporally uniform objects" in one remaining layer is the result (Yuan 1999:3). The consecutive overlay and intersection of all objects at all time steps where a change in one of the objects occurred, will build least common geometries (Yuan 1999:3, see figure 3.10). Each least common geometry which is part of a space-time composite carries the information to which object it belongs at which time. By using the space-time composite method time slice layers for each time step can be constructed by recombining the least common geometries. The practical application of this method is described in chapter 4.4.

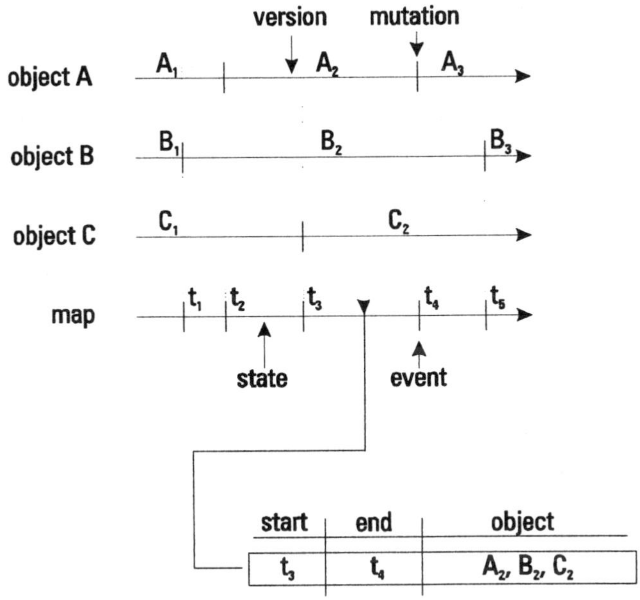

Fig. 3.9 Topology of time (adopted from Langran 1993:33)

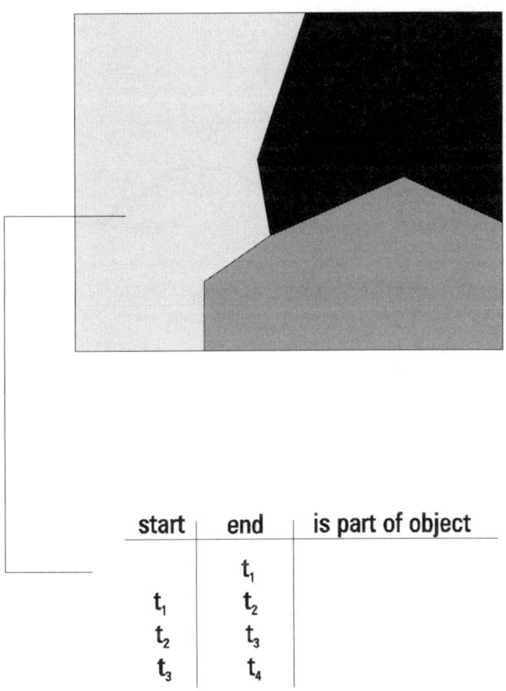

Fig. 3.10 Space-time composite (adopted from Langran 1993:41)

3.2.5
The amendment vector method

The integration of additional events aforementioned concept is in practice a very complex if not impossible task, as the complete structure of the space-time-composite had to be changed. While the least common geometries themselves could easily be updated by intersecting the existing objects with the borders of a changed or emerging new object, the related table storing the information on the recombination had to be rebuild completely. Because all objects are stored as a set of least common geometries valid at a given time, the alteration of one border will create at least one new least common geometry and will therefore effect all related objects.

A space-time composite may also be created by using a system where line segments are used instead of region objects (see chapter 2.3). If an existing line segment has to be altered because of an evolving new border, an additional line segment or a new node in an existing line segment must be inserted (Peuquet 1994). Only one new relation has to be added, stating the fact, that the old line segment P_1 has to be replaced by the new line segments P_2 and P_3 (see figure 3.11). The advantage of this approach is that all objects used to build time slices may be retrieved from the system using relational structures. The performance of such systems will in general exceed region based approaches where least common geometries have to be merged to new objects and the coordinates have to be stored redundantly (see figures 4.27 and 4.29).

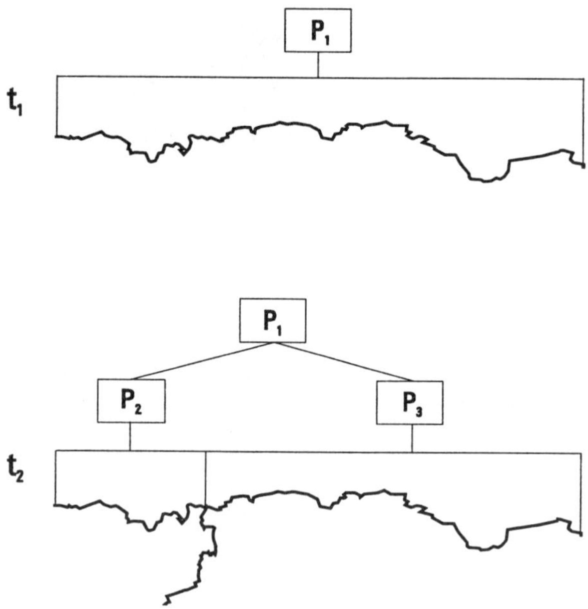

Fig. 3.11 Vector amendment: line segmentation (adopted from Hellwig 1999)

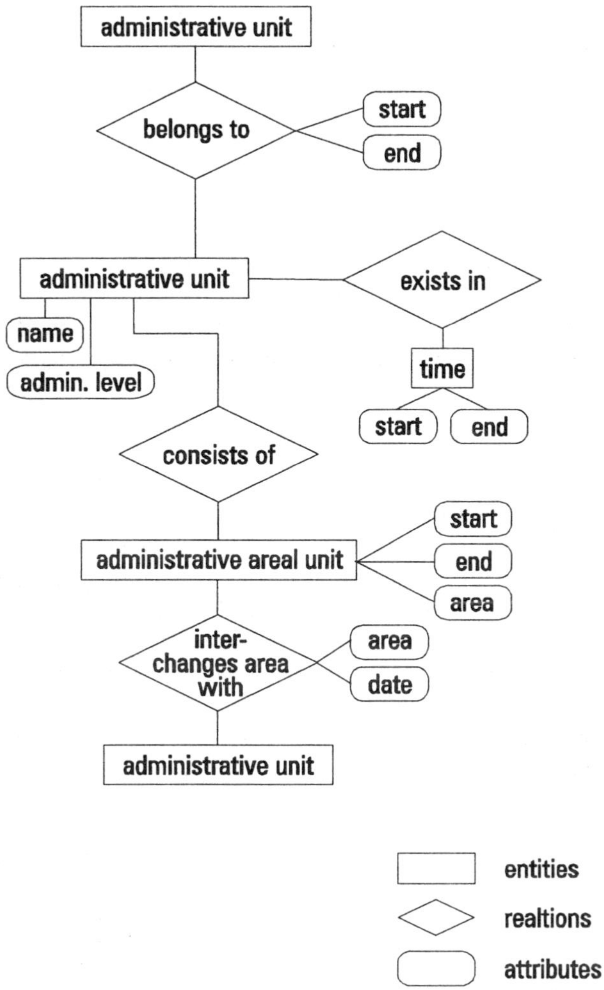

Fig. 3.12 E-R-model of administrative units (adopted from Pierau 1998)

3.2.6 E-R-models of temporal objects

As early as 1978 Basogly & Morrison described a database structure enabling the user to model temporal changes in administrative units. The data structure combined an E-R-model with a simple Arc-Node structure (see chapter 2.3) on the logical level. Another E-R-model was developed bei Pierau (1998) in order to record the territorial structure of Prussia between 1871 and 1945. The basic problem is derived from the fact that territorial or administrative units may only be valid for a certain period of time. They may posses ancestors and successors with differing boundaries. A spatial unit may evolve from either one or the combination

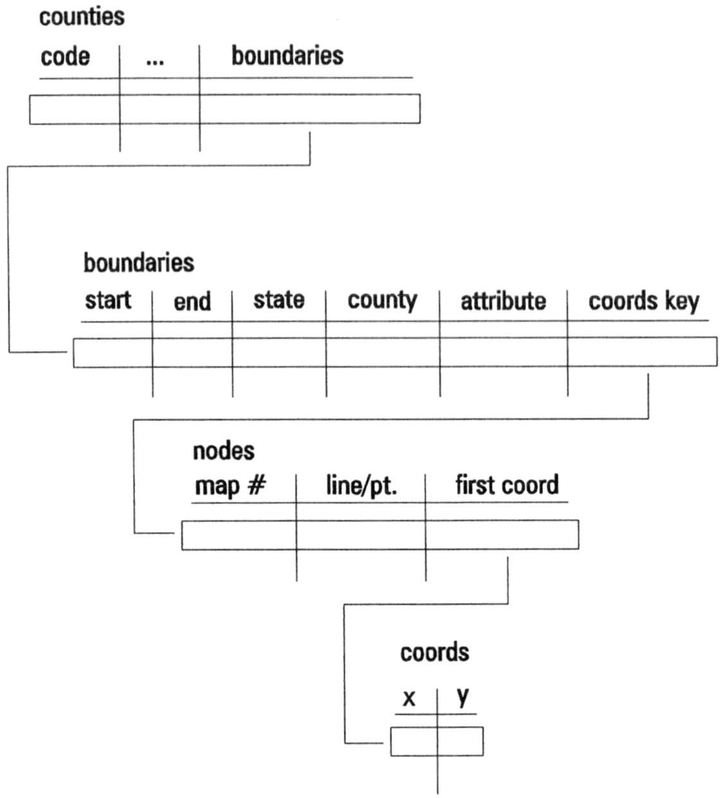

Fig. 3.13 Logical data structure for temporal county boundaries (adopted from Basoglu & Morrison 1978 and Langran 1993:13)

of several spatial units. It may gain or loose area from/to one or several units, It may finally be assimilated in one or several new units. The spatial units are characterized by spatial and temporal attributes as well as by their relations to neighbouring objects. Each object is related to other object categories: a municipality for instance is part of a county or a federal state. Every object is described by attribute data such as census data, which may carry a temporal aspect within the lifespan of the object.

The E-R-model may finally be transferred into a RDBMS in order to collect data on the territorial history. Figure 3.13 demonstrates the extension of the model with a spatial component. The object boundaries consist of line segments (Nodes), the nodes are put together by point data (coordinate pairs). This technique is evaluated in chapter 4.4.

3.2.7
Spatio-temporal object model

Spatio-temporal objects treat the spatial change of entities through time in a three dimensional space, with time as the third dimension (see figure 3.14). Objects between two time sections are represented as volumetric objects. In fact volumetric objects may be thought of as objects represented by an indefinite number of time slices between two changes. As a result spatio-temporal objects cover the time between two changes as a continuum. As it is true for objects conceptualised in time slices, spatio-temporal objects are temporally discrete, they can move through time as well as change their shape. But "none of them are able to portray the concepts about transition, process, or motion." (Yuan 1999:4).

3.3
Event and database time

For the integration of time in an information system two different views on the event have to be considered. "Time is multidimensional. *Valid time* concerns the time a fact was true in reality. [...] *Transaction time* concerns the time the fact was present in the database as stored data" (Snodgrass 1992:25).[28] As shown in figure 3.15 valid or world time can never be later than transaction or database time (as far as events of the real world are investigated). An event cannot be stored in an information system before it occurs in the real world. In a real time system events are stored at the same time they occur, which may be depicted as a line of synchronicity. However, in most cases events are stored in an information system some time after they occur in the real world. If not only real events are stored, valid time can be later than transaction time, e.g. if the results of prognostic models are stored in the database.

The ability to manage both, valid and transaction time is an important feature of a temporal GIS. Objects which are defined by valid and transaction time are called bi-temporal objects.[29] Systems which are able to store both elements of information may be accessed in two ways:

- recover a certain state of the real world for a given date from the most current knowledge,
- recover a certain state of the real world at a given date from the knowledge available at some time step in the past (see figure 3.16).

[28] Italics added by the authors.

[29] Cf. Breunig (1999), Özsoyoglu & Snodgrass (1995), and Worboys (1994;1998) on bi-temporal objects and spatio-temporal complexes.

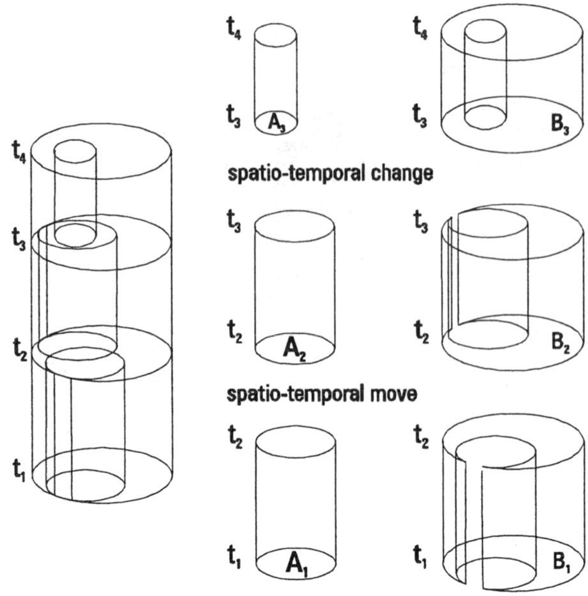

Fig. 3.14 Spatio-temporal objects (adopted from Worboys 1992 and Yuan 1999:5)

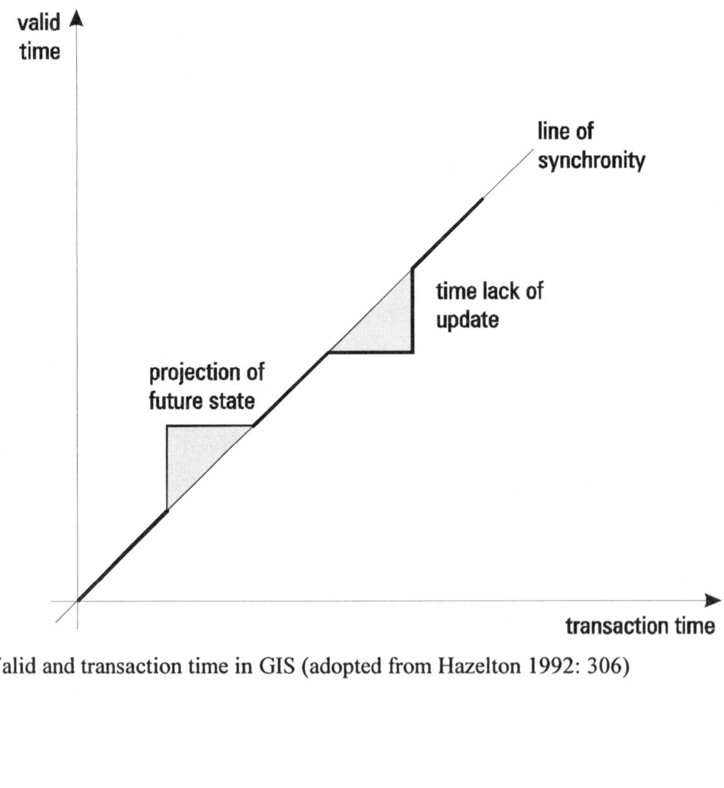

Fig. 3.15 Valid and transaction time in GIS (adopted from Hazelton 1992: 306)

World = Database	What do we know of the world on this date, as of the same date?
World < Database	To the best of our knowledge as of a given date, how did the world appear on a given past date?
World > Database	Based on our knowledge as of a given date, how would we expect the world to appear at a later date?

Fig. 3.16 Queries made possible by tracing both world and database time (Langran 1989: 219)

3.4 Version Management

Spatio-temporal objects are limited to a given temporal validity. When new information is assigned, the database has to be changed, in order to add the new data. In such a case the database may have to be rebuild completely for storing the new state, which causes a great amount of redundancy. Alternatively the database might be time stamped in order to indicate the temporal validity of its objects (Kemp & Kowalczyk 1994:92). According to Langran (1989:220-5) three alternative approaches can be used to update a relational database (see figures 3.17-18):

- *Relation-level versioning:* This method applies a roll back concept. For each event a new snapshot is created indicating the time from which the current state is valid. The kind of changes may be identified, e.g. by using status information like added, altered, deleted, updated, restored. Elements not valid at the moment may be kept inside the system in case they need to be restored later.
- *Tuple-level versioning:* Only the changes are stored together with the according database and world time. Time slices for a given transaction time can be derived, with negative effects on the performance even for current time step data.
- *Attribute versioning:* The changes occurring in the temporal database are stored at the attribute level. Compared with the tuple-level versioning, changes in entities and in attributes are time stamped separately.

The ability of versioning a database is essential for all spatio-temporal applications, where knowledge of the real world is uncertain or changing through time. Wachowicz (1999:58) gives an example of changing steps in the life span of a boundary. Applied to a legal parcel these stages could be described as follows:

- An existing border has to be changed because of a legal change. The actual boundary is no longer valid, but may remain in use for some time before a new boundary is inserted into the system and made available to the public.
- With the physical demarcation a provisional map is created. This is a draft version of the now valid boundary. This boundary may not yet be used, because quality check has not been completed.
- After the quality control is passed successfully and errors are fixed, the map can now be used as a final map replacing the old one.

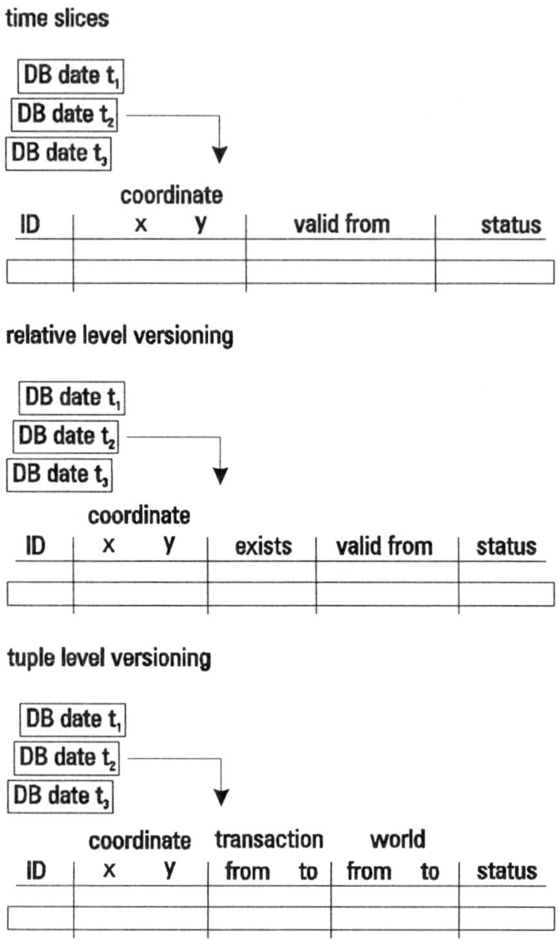

Fig. 3.17 Version management (I) (adopted from Langran 1989:220-225)

The example shows, that the system must be able to store at least two different states and versions of the border in the GIS. Please note, that the valid time and the database time of both versions differ.

During the process of updating a border, different statuses, e.g. actual valid border, draft of new border, revised edition of new border and finally the approved new border are used. According to the status the border may not be edited (e.g. in the case of valid border or border to be published) or they may be edited (like draft borders) (Ramsch & Sosna 1997). At the end there must be a check on consistency, so that not more than one version is accessible to the public. In the given example the status of the actual border has to be switched to "old" when the new map is made available to the public and the status therefore switched to actual. This can be made either explicitly or by adding a final time stamp to the database indicating that the border is no longer valid.

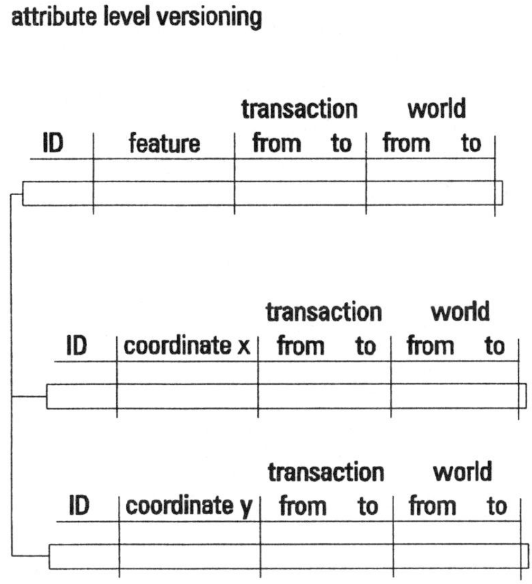

Figure 3.18 Version management (II) (adopted from Langran 1989:220-225)

3.5 Types of temporal databases[30]

Temporality can be implemented in databases in different ways (Armstrong 1988: 982-983, Wachowicz 1999:40, see figure 3.19):

- *Static databases:* store only the information currently valid. Old data will be overwritten by emerging new data.
- *Static rollback approaches* store the old data every time new data is added to the database. The resulting time slices allow a rollback view on the old data. Unfortunately all the data has to be stored redundantly, so the approach is not effective in all cases where only minor changes occur or updates are made frequently. This approach is applied in most update processes for data security reasons. In this case only the transaction time is stored, as the system covers only the time when the event is added to the system not the time the event occurred.
- *Historical databases* store the valid state at a specific time step. This approach uses the valid time an event occurred. Historical databases are able to store both, time slices only valid for specific time steps or the validity of objects between events. In both cases changes to the data can be made at any time, but the database will only be time stamped with the valid time.

[30] Cf. Özsoyoglu & Snodgrass (1995), Kemp & Kowalczyk (1994), Snodgrass (1998), and Wachowicz (1999: 42) for an evaluation of temporal databases and query languages.

- *Temporal databases* store bi-temporal data. For each object real time and transaction time is stored. With this approach it is not only possible to represent the current state at a given time step but also which state was known to the system at a given time step. This is important in all cases where the records of a temporal process cannot be recorded in a chronological way.
  This could be valuable in all cases, where we try to understand, how our knowledge of the state of affairs changed or improved. It may be necessary to recover the knowledge available at a certain date to analyse decisions made based on this information.
  If different sources are used for spatio-temporal objects not exactly defined, the ability to store more than one transaction time may be adopted to cover different sources.[31] In this way different views on one world representing different levels of knowledge existing at different time steps can be managed and analysed. It is also possible to store different alternative projections of the future, applying different assumptions (see figure 3.3).

	no support for transaction time	support for transaction time
no support for valid time	static	rollback
support for valid time	historic	bi-temporal

Figure 3. 19 Temporal dimensions supported by system types (Worboys 1994:28)

Such databases are still predominantly built using relational database technology. There has been a considerable amount of research and developing on relational database and retrieval systems which support time in one way or the other. Common to all these approaches is, that the temporal information needed to define an object is stored in a database separated from the storage of the geometry data itself. In a layer GIS the only way to time stamp individual geometry objects instead of complete layers is to use a space-time composite. In chapter 4.4 an approach is presented how to use least common geometries in order to represent space-time composites. They are time stamped in an additional table which allows also to link the object within its live span to a temporal attribute database. However, the pseudo object approach based on the amendment of line segments, which is discussed in chapters 2.3 and 4.4 eliminates some disadvantages of the space-time composite approach using least common geometries as a basis. Using an ER-model for the spatial representation of the objects and their chronology, the redundancy may be reduced and incremental changes stored separately.

[31] Cf. Worboys (1998)

As described in chapter 3.2 the ER-model approach (see figure 3.17) may be utilised to allocate constant identities to objects which spatially changing (Kemp & Kowalczyk 1994; Langran 1989). Objects may be stored with:

- Predecessors and successors and the respective reason of change.
- The valid and transaction time for each instance of the object.
- The spatial representation of each instance of the object, either by point, line and region object or raster cell.
- The hierarchical position of an object and the respective relations to other objects and object classes.
- The ID for the geo-coding of the attribute data valid for each instance of the object.

Finally, a completely object-oriented approach is at the moment only addressed in academic discussion (see chapter 2.3). The improvement of the pseudo-object approach with object-oriented databases and programming languages will be a task for the future.

4 Implementation of time in GIS

> "The implementation of a GIS is where technology and people meet."
>
> (Stan Aronoff 1989: 249)

GIS projects often fail even before they have started. Careful planning is needed in order to avoid common pitfalls and errors. This chapter is intensively concerned with the construction and implementation of a time-integrative GIS. Special attention is given to various aspects of data acquisition, storage and processing.

4.1 Implementation problems

In general the implementation process of a GIS consists of several stages, where decisions have to be made, which are critical for the success and for the sustainability of the project (see figure 4.1). The implementation process as a whole is closely related to the conceptualising of the real world in a GIS and the successive stages of building a GIS described in the previous chapters of this book. Decisions made during the initial phase of a GIS project define the resulting functionality and operability of the system. So, right from the beginning, where first decisions have to be made and money has to be spent, the complete implementation process has to be overlooked and evolving interdependencies and constraints must be considered. In practice an unprofessional project management could seriously threaten the success of the implementation of a GIS.

In the case of developing a spatio-temporal GIS the danger of failure is even more likely because at current there are no off-the-shelf solutions. Furthermore there is hardly any expertise available for this particular field of application outside research projects dealing with the topic.

A second source of problems arises from the great success of standard GIS packages in recent years: the diffusion and widespread usage of GIS beyond the traditional fields of application in geography and closely related fields like urban and regional planning or more technical fields like facility management. More and more people not familiar with the basic concepts of spatial research and analysis, without expertise in database or GIS design feel the need to use the advantages of GIS for their work. In such cases the decision for the implementation of a GIS is often made during an existing project, often at a point near the end of a project or when given project goals cannot be reached using other conventional techniques.

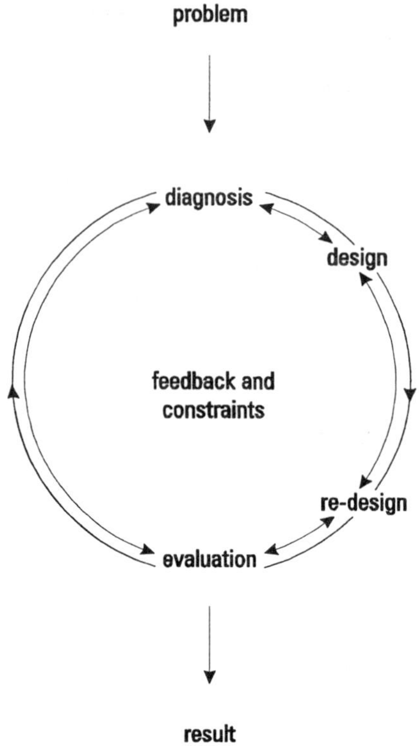

Fig. 4.1 Project design (adopted from Hendrik & Ottens 1997:19)

In addition, temporal data is mostly collected, stored and used without a spatial context. Spatio-temporal relationships have to be (re-) build in order to meet the characteristics of the existing data and the requirements of the research or work process already established. This very often leads to a sub-optimal conceptualising and GIS design, resulting in severe restrictions of the possible spatio-temporal analysis methods.[32] Data collection and processing, database design etc. are often already completed and no sufficient money or manpower is left for building the GIS.

As a matter of fact there is a growing demand of low cost, low tech and ad hoc approaches. The integration of time is still a desiderate in GIS development (Behr 1998: 32).[33] The suitability of the available GIS packages for representing spatio-temporal data varies. In general, object oriented GIS provide the system developer with better and more versatile abilities to deal with the demands of spatio-

[32] There are three ways to implement a GIS: first, oriented on the acquisition of the data, second, oriented on the queries, third, oriented on the required output. See Molenaar (1998:3) for a detailed discussion and further references.

[33] "The temporal validity of attribute values and geometries is currently not sufficiently supported by data storage systems and GIS (Behr 1998:32)." Translation by the authors.

temporal data (see chapter 2.3). Unfortunately the most common and in terms of cost and training most applicable GIS packages are based on a traditional layer concept. Therefore the following sections deal with the most serious implementation problems and present a selected set of approaches that may be implemented easily with standard layer-oriented GIS packages.

4.2 Project design

The implementation of a GIS requires consecutive steps of action.[34] Decisions made at the beginning will influence the outcome of the whole project. They possibly cannot be changed afterwards without large costs or the necessity of a complete re-design of the project. The implementation of a GIS has not only consequences for those involved in the work with the new GIS itself. It may influence the work process or procedures of other people as well. It may also become necessary to alter existing work processes to adopt them to the new structures created by the implementation of the GIS. People involved in such changes have to be contacted and informed, the changes must be documented. The necessary changes in hard- and software demand resources and training of the people affected in order to cope with the new requirements. The impact of the implementation of the new GIS has to be checked carefully. Consequences may fall far beyond the competence of the GIS implementation group. Lastly legal aspects have to be taken into consideration. The amount of powerful GIS analysis techniques may affect an established privacy protection policy or legal requirements restricting the use of individual data.

Besides this administrative view also technical aspects must be planed ahead carefully. A survey of available software packages and techniques, references of the expertise of external personnel or firms, plans for data security and safety recovery are standard within a professional project planning process. If necessary the advice of the computer service department or external consultants should be included. The development of data structures and the adaptation of existing geometries and attribute data for a GIS as well as the building of tools and a individual user interface can be far more expensive than the costs for hard- and software. Therefore sustainability is one of the most important criteria for the selection among the numerous hard- and software solutions. The decision for a hardware environment should ensure long term support and maintenance as well as scalability to adopt the system to growing performance needs in the future. This applies also to the selection of software packages or the decision to develop parts or the complete software, tools, and user interfaces from scratch.

Another crucial point in project planning is the control of financial resources, manpower and time. The estimation of time needed to complete single project steps and the complete project may be almost impossible, if new techniques and programs have to be developed. Not calculable risks may arise, if the project tar-

[34] Cf. e.g. Hendrik & Ottens (1997:19).

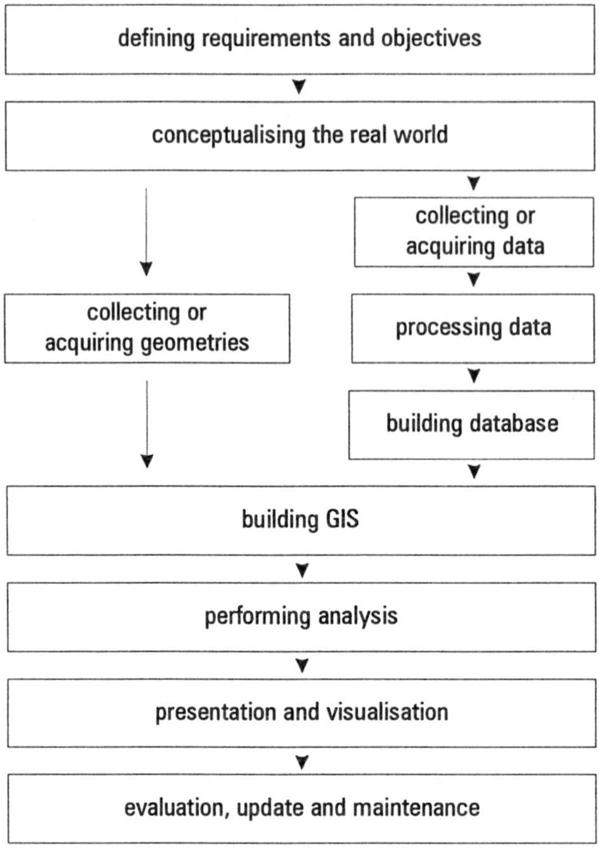

Fig. 4.2 Project implementation: top down approach

get is changed while the project is running or inconsistent goals have to be integrated into an already running project to deal with emerging new requirements. Most projects depend on work in a project group. People not directly involved with the implementation of the GIS may have different priorities and interests. They may therefore not be able or sometimes not willing to contribute to the work of the project group or to complete work for the project group at time.

At the end of the implementation process precautions have to be taken to secure the long time operability. The organisation of the update process of geometries and data is especially important in the case of spatio-temporal GIS, where processes in progress are most interesting. The final step is the evaluation of the project results. Success in terms of functionality, user acceptance, time and cost savings etc. should be measured and used for the design of further project stages ot to fix requirements not yet met.

The importance of a professional project planning for the implementation of a GIS cannot be overestimated. The examples so far should give an impression of the character of problems which could endanger the success of the project. Project planning is a professional field of itself, in this chapter only some important aspects could be compiled.[35]

Basically, the project planning can be differentiated into two opposing approaches: top down and bottom up. Both are characterized by a number of advantages and disadvantages which will be explained in the subsequent sections.

4.2.1. Top down approach

The top down approach fits fine with the process of conceptualisation described in chapter 2 and is similar to the process of empirical research applied in the social sciences (see figure 4.2). First of all, goals and requirements must be defined. In the second stage the conceptualising of the real world takes place, based on theoretical insights on the nature of the problems, processes, spatial- or spatio-temporal structures which should be included into the model and the analysis functionalities needed for the project. The following steps are of a more technical nature and deal with the identification of the necessary geometries and data, the capturing, processing and storage of this information. In this context a logical and a physical data model have to be developed or chosen from the available database and GIS packages. Again, this decision is of crucial importance and directly linked to the goals defined in the first instance and the corresponding analysis and modelling tools. When the GIS is finally running the main concern lays on the analysis of the data, the visualisation of the results, and the update and maintenance of the system.

The top down approach demands a complete overview of the system requirements and goals and a detailed survey of possible obstacles and constraints. This centralised approach usually concentrates the responsibility at the top level. Know how and the making of decisions remain with the project group which has to ensure the collection of all information needed for the implementation in advance as well as the coordination of the project group. The advantage of this straight planning process is, that functions, operability, costs and duration of the implementation process may be optimised. The integration of the GIS in existing frameworks and alterations of existing work processes should be made in such a way that the advantages of the new GIS are maximized. Where sufficient know how lacks in the respective departments the top down approach is the only way to introduce a GIS without changing the responsible personal. However, major changes into existing structures and working processes are inevitable in order to provide an environment where a GIS can be conveniently introduced.

A serious disadvantage results from the centralising of decision making. Special knowledge acquired during long practise may be overlooked and personal know how not adequately utilized for the forthcoming of the project. A top down

[35] For a more detailed discussion of this topic in general and applied examples in GIS please refer to Aronoff (1989), Burrough (1986:167-176), Behr (1998), Huxhold & Levinsohn (1995), Inmon (1986) and Klemmer & Spranz (1997).

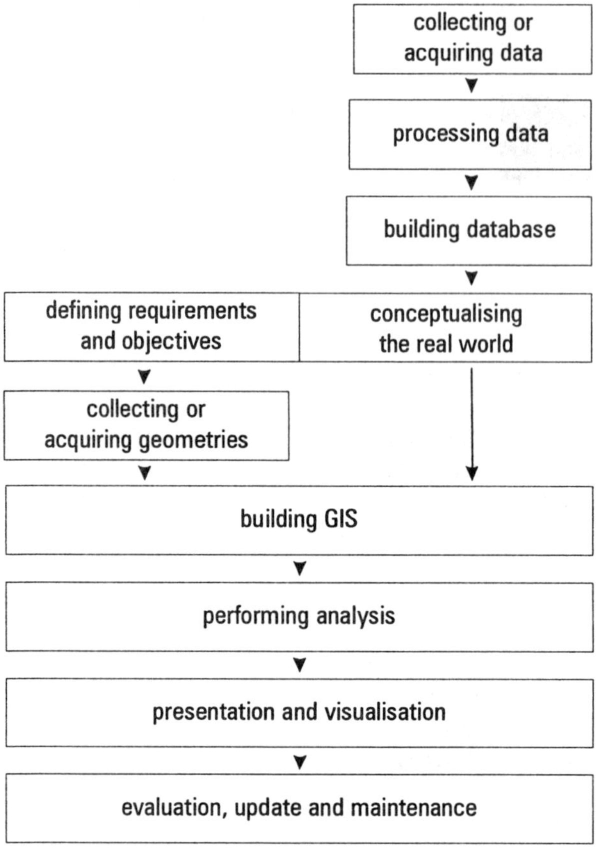

Fig. 4.3 Project implementation: bottom up approach

approach could best be utilized for the implementation of a GIS in a complex, heterogeneous environment, where a considerable number of people use standardised services provided by the GIS maintained by a centralised GIS-group. Thus, the implementation project must administer a considerable amount of resources over a period of time because of the costs of the centralized project planning activities.

4.2.2 Bottom up approach

The bottom up approach is based on the awareness of individuals that they need GIS to fulfil their tasks or that they could use GIS to improve their research results or the services they provide (see Figure 4.3). In contrast to the process of conceptualising a GIS described in chapter 2 the bottom up approach starts somewhere at the end of this process. In most cases data is already available, the processes of data acquisition and processing are worked out and databases are developed. This set of data represents a large amount of money invested before the implementation process of the GIS starts. The ad hoc decision to use a GIS for the processing of data, to utilize the analytic functionality of a GIS, or to provide new services means that there is often no or no sufficient additional funding for the implementation in the first instance. Thus, the GIS has to be build using as much existing resources as possible. Often it is also impossible to employ specially trained staff for the implementation so that the project has to rely on the expertise of the employees already involved.

On the whole the bottom up approach suffers from a lack of money and manpower which causes sub-optimal results. In most cases it is not possible to develop special software tools to meet special user requirements or project goals. Therefore the potential benefit of GIS cannot be maximized (e.g. by developing user defined interfaces or standardizing common procedures). Off-the-shelf solutions have to be used whether they are appropriate to the problem that has to be solved or not. Research projects at universities often have no choice at all and have to cope with the software packages made available by the university's computer department.

This framework, in which the implementation has to take place, sometimes dramatically cuts the number of options to build a GIS. The existence of pre-structured data, which is often the most expensive part of the whole GIS, has decisive influence on the conceptualisation of the real world. Geometries which are used to spatially represent the existing data have to fit the structure of the given data. Also the building of logical and physical data structures are pre-defined by the data itself. In this case, as a matter of fact, analysis and visualisation methods are not subject to proper project planning at all. The outcome of the project is a mere result of cumulative constraints. In such a situation it is almost impossible to integrate the GIS into global working processes and to coordinate the implementation with adjacent processes or the work of neighbouring departments.

The results of the bottom up approach are often insular solutions without long term sustainability. Investments in hard- or software are often made without the guarantee of reasonable support over a sufficient period of time. Under such conditions a detailed project planning is almost out of the question. The use of resources, manpower and time is more likely to be decided by the circumstances then by the project planning.

Nevertheless, the results which can be achieved are highly determined by conditions often not to be influenced by the project group itself. Therefore an evaluation process should be performed. Even if not formally applied this could produce useful information on how to improve the use of the GIS or the GIS itself.

The bottom up approach should only be used when planning is not possible or when other more planned approaches are not applicable. On the other hand bottom up approaches facilitate the start of a GIS implementation by using small amounts of resources and mobilizing knowledge available in a project group. The results of such efforts could be used to show the advantages of GIS to raise additional funds or to increase the acceptance of a GIS implementation. The information on requirements and possible fields of application collected in a first ad hoc stage of implementation could be used in a second stage to develop and implement a more sophisticated solution.

Table 4.1 Usage of GIS components (Bill & Fritsch 1994:4)

Hardware	3-5 years
Software	7-15 years
Data stock	25-70 years

	user develops system	software packages	GIS services
dependence on supplier	very low	high	almost complete
time until start	long	relatively short	short to long
initial costs	low	moderate	relatively high
risk of failure	high	relatively low	minimal
customizing and adaption to requirements	complete	moderate	varies
technical skills of project group	extremely high	moderate	relatively low
utilizing existing resources	high	moderate	very low

Fig. 4.4 Buying or building a GIS (adopted from Aronoff 1989:263)

4.2.3
Acquiring the technology[36]

Acquiring hard- and software are necessary parts of the GIS implementation process. The decision which hard- and software to use and whether the software should be developed from scratch or bought from the shelf is important. The perspective of this decision reaches from some years for the hardware till decades for the data stock (see Table 4.1). "Depending on an organisation's resources and budgeting needs the hardware and software may be purchased before, after or during the database development" (Aronoff 1989: 262f). Buying the hard- and software at the beginning of the implementation process has some advantages. First the GIS could be used to check the interoperability of the data collected and processed. Expensive adaptations could be avoided in this way. The GIS could also be used to train the staff, so that sufficient experience is available at the point when the work with the GIS should begin. On the other hand there is probably a relatively long time to work on data acquisition and processing before the GIS could be used, the investment in the GIS will not be used efficiently during this period. The rapidly increasing technological development results in falling prices for more powerful and advanced hard- and software so that "it is usually best to delay the system purchase as long as possible" (Aronoff 1989: 263).

Starting with the development of the database and buying the GIS later could lead to additional costs for converting the database to fit the finally chosen GIS. However, one advantage is that one can buy the most up to date GIS technology. Aronoff (1989:264) suggests to make the decision for a suitable hard- and software solution before starting with the building of the database. If necessary a pre-contract with the vendor could be negotiated so that support and training is available. The delivery or installation of the then latest version could be timed in accordance with the development of the implementation process. If possible the maintenance and updates should also be negotiated right from the beginning to ensure that the follow up costs may be estimated for the project planning.

Another question is, which parts of the complete project should be kept in-house and which parts should be bought from vendors or GIS service providers. The decision depends on several aspect which are shown in the matrix of figure 4.4 The building of the complete system in-house requires advanced technical skills of the project group. The risk of failure is high, and the time span before the system is ready for use is much longer. On the other hand the initial costs are low and a lot of existing resources, like hardware, software, manpower and specific knowledge may be utilized for developing the system; there is almost no dependence on supplier nor the need to give internal information and knowledge to external people. This method is very flexible in terms of functionality and integration into existing structures.

Buying the software is nowadays the most common way of building a GIS. The time span before the GIS can be used is relatively short, the risk of failure also

[36] This chapter is adopted from Aronoff (1989:249-278). See the affiliated references and literature for a more detailed coverage of this topic.

relatively low, the expertise requirements of the project group and the initial costs are moderate. Disadvantages are the dependence on a supplier and the relatively inflexible opportunities to adopt the GIS to specific needs; the use of existing resources is limited. This approach is the most versatile and suitable for most purposes.

A GIS could also be build using external GIS service vendors. In this case there is no need for specific knowledge of the project group. The period before the GIS can be used depends only on the complexity of the system. The risk of failure is minimal, but as a result the initial investments are high. The possibilities of customizing the system depend on the service provided by the vendor and the costs charged for such services. The dependence on the vendor is on the other hand almost absolute and the use of existing resources is limited. Internal information and knowledge have to be shared with the service provider which might be a problem in terms of security and confidentiality. This approach is advisable if complex problems have to be solved or sufficient know how is not available in-house.

4.3 Implementation constraints

In most cases the implementation process is restricted by several restraints. The following overview summarizes once more the main points:

- The requirements and goals which should be reached by implementing a GIS are not clearly defined. It may not be possible to integrate antagonistic goals in one single GIS design.
- No project planning is applied. The needed resources, manpower and time are not estimated and run out of control.
- The project suffers from under provision of relevant resources like money and manpower. As a consequence there is no acceptable level of knowledge and software available.
- There is not enough time for the development of a GIS. Ad hoc solutions have to be adopted.
- The existing working processes and frameworks are not compatible with the requirements of the implementation of a GIS. The coordination of the implementation with existing structures is insufficient.
- The knowledge of the existing personal is inadequate, there is not enough training available, or the acceptance of the new technology is low.
- Attribute data, geometries and information on spatio-temporal structures could not be acquired to meet the requests of the GIS, e.g. because of copyright reasons or high costs for data collection and processing and for digitising suitable maps.
- The geometries and data are used as they are. No conceptualisation takes place, spatial representation of data is incorrect. Inadequate analysis and visualization techniques are applied.

People responsible for the implementation of a GIS have to be aware of this constraints and handle them in a responsible way. The last two points are directly related to the building of a spatio-temporal GIS. In the following section 4.3.1 data sources as possible constraints of GIS implementation are discussed. Chapter 4.4 deals with the problem of building a spatio-temporal GIS with inadequate sources.

4.3.1 Data sources

For most spatio-temporal GIS the collection of data is the most expensive part of the implementation process. Problems evolving during the collection and processing of the data may be of different origin. For already existing databases the question of compatibility is important, varying database formats have to be converted. Often additional processing is necessary to enable the GIS to work with the data.

This also applies to data bought from different vendors or acquired from administrative bodies with different data standards. In some cases no metadata is available, so that there is no information about the sources, scale, precision, date and definition of the variables, which could lead to severe problems ending with a complete renunciation of such insecure sources.

The fact that there must be one common type of information to geo-code the attribute data with the corresponding geometries often leads to expensive manual recoding of the attribute databases. For spatio-temporal GIS a third kind of information is necessary. This is hard to obtain, because it is commonly stored in a way which cannot be applied in a GIS easily. For the geo-coding of spatio-temporal data it is necessary to correlate geometries with corresponding attribute data whereas in a spatio-temporal environment each spatial and attribute information used has a temporal validity which has to be added to the geo-coding process (see figure 4.11).

4.3.1.1 Attribute data

Time and temporal information may be part of attribute data in different forms:

- Point in time of events occurring.
- Periods of time related to events.
- Attribute values at a single time section.
- Attribute values at consecutive time sections (time slices/time series).
- Changes during a period of time or between two time sections.

Attribute data with the point in time of an event might for example be the information on the year of a battle in history or the date of relicts in archaeology. Such information is mostly collected from archives or external sources, but might also be the result of field research. It is also possible to detect the movement of an object over time by using information of this object for more than one date.

used to match the dataset with an already given geometry object. The same structure may be used to store periods instead of the date of an event. Periods could be the age of buildings in urban planning, the age of relicts in archaeology, or travel time in logistics.

Attribute data for a single time section are most common. Each statistical survey or census gives information on variables for one defined time section. A spreadsheet or database containing such data can be geo-coded without any effort to corresponding geometry objects for the respective time section. This might be called a mono temporal GIS, the least complex application of time in GIS.

The attribute data for consecutive time sections may be differentiated according to their covering of temporal changes. The ability to geo-code geometry objects with the attribute data depends on this characteristic. The first possibility to store data for two or more time sections is to build time slices. In this case no information on the temporal relationship between objects at different time slices is needed. For each time slice a corresponding set of geometry objects is required. By contrast, information on the temporal relationship of objects is essential for the second possibility. For each object at one time slice a corresponding object at another time slice must exist. All corresponding objects may be geo-coded with one geometry object, valid for all times slices. From such data time series can be derived. A special case are time series from one source covering specific periods of time. For such data obviously only one geometry is needed for the geo-coding process.

In many cases the development between two time sections is of special interest. Such changes may be derived and analysed by calculating new variables from the variables for two time sections. Such variables could e.g. be the decrease or increase in percent during a period of time. This requires data with temporal relationships as described above. Even if such information is not available, some variables covering changes are also frequently calculated and provided for time slices and could be used with the corresponding time slice geometry data.

The storage of such temporal data can be organized in different ways. There are four types of information which define the data matrix of an attribute data collection:

- the definition of the variable (including the measure if applicable),
- the temporal validity,
- the spatial coverage, and
- the variable value.

Most data is obtained or collected in form of a table or spreadsheet which may be stored easily in a relational database (see chapter 2.3). Initially there is no need to process or convert the data to use it in a GIS as long as the format is readable by the GIS. In practice there are several reasons to compile all attribute data, or at least large parts with corresponding content or source, in one consolidated database. This database should provide a live link to the GIS (e.g. Microsoft OLE), so that changes in the database are automatically transferred to the GIS. This could be done in two ways:

- The first option are conventional tables in a relational database management system where variables and data sets are cross tabulated (see figure 4.5). Eventually large data sets may be separated into relational tables to prevent redundant storage. In tables only two dimensions can be represented using the rows for one dimension and the columns for the second dimension. The variable is stored in the fields of the table. One table can contain only one variable: If the rows are used to identify respective territorial units, the column may be used for one variable at different time steps or different variables at one time step. Thus, the structure of such a data matrix is not versatile. The possible analyses are restricted by the structure of the data tables and a redesign of the table in order to allow additional analyses is time consuming and inconvenient.
- A more flexible data storage can be applied by using a standard relational database management system and an alternative design of the table (see figure 4.6). Each information described above: variable, temporal validity, spatial reference, and variable value supplemented by additional meta information like sources, comment on the data etc. may be stored in columns of a data table. For each information an additional row is added to the table. To reduce the redundant storage of data, e.g. of meta data, relational tables could also be applied in this case.

  With this method the number of data sets is sometimes several times larger than using conventional tables. For example, a data matrix of 10 rows (cases) versus 10 columns (variables) for 10 tables (time steps) results in one table with 4 columns (case, variable, variable value, time step) and 1.000 rows.

  The performance of modern RDBMS allows the storage of huge tables, so that this technique could be used with nearly every RDBMS, even on a PC. From this data matrix tables can be derived using simple cross tabulation functions of the RDBMS with variable column and row headings. For a convenient integration with the GIS these tables may be stored separately.

  In cases where the data matrix cannot be overviewed during the collection process this data structure is highly recommended, as it can deal with not clearly defined territorial representations of the data and objectives of analysis.

4.3.1.2 Geometry data

Geometry data may be obtained from data vendors, statistical or survey offices, or generated by manually digitising suitable maps. The temporal dimension of such geometry data is usually that of a snap shot, representing only one time slice. The temporal validity of the data has to be documented in a set of metadata or captured from the digitised maps and additional sources if necessary.

All common GIS use data structures neither able to cover more than one time slice in one layer nor multi-temporal objects. Temporal dimensions can only be added by using additional layers for succeeding time slices in which objects have

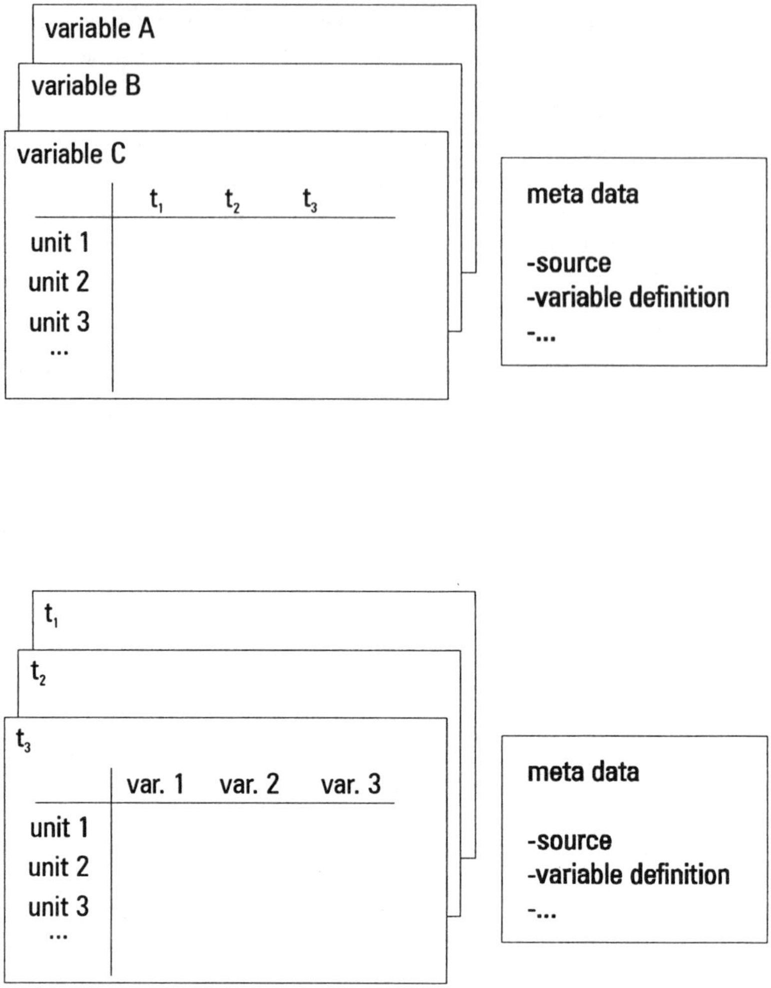

Fig. 4.5 Storage of spatio-temporal data in three dimensional RDBMS tables

to be stored redundantly. Furthermore, the temporal validity of such time slices is often only indicated for one specific date or year. Nevertheless, the time slice can be valid for a certain period of time between two changes. Information on territorial and areal unit changes taken from external sources (see chapter 4.3.1.3) can be utilized to find out the necessary facts to determine the period of temporal validity of a single time step geometry layer.

The organization of the geometries is predetermined by the data model of the respective GIS. In most cases this model will be a layer concept, where objects representing one class of entities are stored in one layer (see chapter 2.3).

The logical and physical data structures utilized in different GIS packages are diverse. They may be differentiated according to the accuracy, the spatial reference system used to store the spatial information etc. Most packages provide only limited abilities to read and store spatial data in other data formats or projections. Compared with common databases and spreadsheet programs, which support a lot of data formats and enable the user to convert data from one format to another without loss of information, converting spatial data files for GIS is much more complicated. The user often needs a great deal of technical experience to develop special interfaces, or expensive converting tools must be acquired. Because of different data structures it is often not possible to convert data between two formats without a loss of information. For that reason digitising and acquisition of spatial data must take this potential incompatibility into consideration: data should possibly be digitised or bought in the same data format which is required by the prospective use.

Compared to attribute data the usage of geometry data is a more complex field. The definition of a suitable database structure and the collection of data from secondary sources are problems which can be solved easily. Conventional techniques can be used, which are commonly applied by most researchers and database administrators. Technological or methodological problems are no constraints for the building of suitable datasets.

Geometry data is more complex and hindrances in terms of technology and methodology are common. In most cases they are solved by trial and error approaches or by adopting conceptual and logical models from already established and successful applications, whether such approaches are appropriate or not.

In chapter 2 the ontological problems of conceptualising real world phenomena using vector objects was discussed. These problems must be re-evaluated in terms of usability of spatial data sets at this point. In most cases theoretical approaches for the conceptualising of the real world can only be applied in a very limited way. Until today, the most common sources of digital spatial vector data sets are conventional paper maps. Most of the digital vector data sets which can be bought from different vendors, were also manually digitised from paper maps. Although digitally produced data sets can be organised in a completely different way, the production of such spatial databases is in most cases similar to the preparation of analogue maps in the end. They will therefore not be treated separately in the further discussion.

When choosing paper maps for digitising own datasets or buying digital maps from vendors, the user should look carefully how such maps are produced and asset their quality. The ability to use this data for maps depends on these properties. Three major points have to be considered:

- In which scale is the map drawn or digitised? Which projection was used?
- What is the content of the map and how is the real world represented by vector objects?
- Which temporal information is included or provided with the map?

The map scale is always important, as the representation of real world entities in a map depends on the scale of the base map. Figure 4.7 shows the representation of a city in different levels of scale. In large scale maps the representation may deal

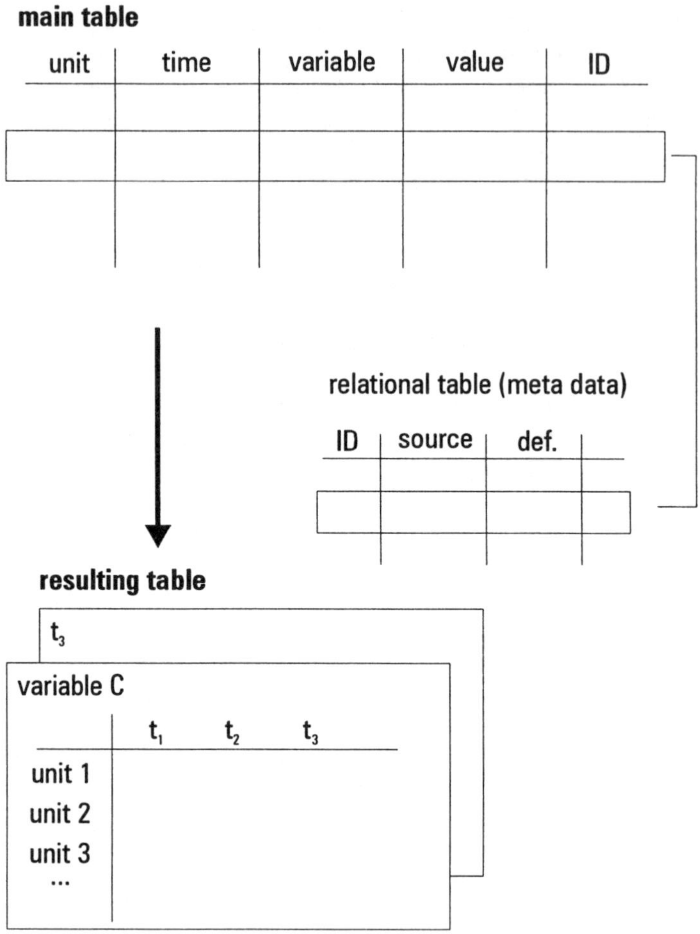

Fig. 4.6 Storage of spatio-temporal data in a multi dimensional RDBMS table

with different aspects of the entity city. Besides the administrative boundary e.g. the build up area might be displayed. In the next smaller level of scale the city has to be reduced to less information. In the given example only the administrative border remains. Administrative boundaries are one of the most commonly used region objects because of their unproblematic availability for large areas. The problems and limitations of administrative boundaries in terms of conceptualising the real world were discussed in chapter 2. This example demonstrates the obvious: with smaller scale less information can be stored and presented. Another step further the same border can no longer be printed in the same accuracy as before. The border was generalized to reduce the amount of information. As a result the resolution of the border was decreased. Now, there is no longer a direct reference between the geographical location of the border in the real world and the representation of the border in the map. The level of abstraction of the region object as a model of the real world entity city is now increased. In the last instance the generalisation results in a symbolic representation of the city. We know such symbol representations of locations from atlas or street maps. In small scales it is often no longer possible to use region objects to represent areal units in the real world. With this generalization the representation of the city is reduced to an abstract concept of location. In practice the position can be located at the centre of the city, e.g. using the central business district, a central place or building like a church to locate the symbol. On the other hand it may also be derived mathematically from a region object by different types of interpolation. In this case we lose not only the important information, that the city is an areal object we are also no longer able to calculate and address this respective area in the GIS. Whether such point representations are geographically correct or not depends on the method, by which the points are determined. Generalization from areal units to points should only be accepted as long as the resolution provided by the possible scale is smaller than the occurring spatial inaccuracy.

Standard GIS provide several functions with which generalization and the transformation from one object type into another can be performed. Besides the high cost for the digitisation or acquisition of large scale maps, there is no need to use small scale maps in a GIS in the first place. Data storage is in most cases no longer a limiting factor and data analysis will generally profit from a higher degree of resolution and accuracy. For the presentation, if necessary, the resolution and scale can be reduced to fit the possibilities of the output device.

The second important information displayed on maps or represented by digital spatial data is the content of the map. In most cases topographic maps are used for digitising, whether done by oneself or bought from a vendor. A first overview of the content of the map can be derived by reading the map legend. Fortunately the production of topographic maps is highly standardized, the production process defined in detail. The rules applied to select and draw real world entities are clearly described. This applies as well to the generalization process for maps produced by scale reduction from other maps. Digital maps always consist of the data set itself and additional meta data describing e.g. the source map, the resolution and the organisation of the objects in layer.

In the real world a lot of phenomena are of continuous character. Such continual changes must be transformed into discrete steps, so that they can be represented by vector objects (see figure 4.8). This is always a problem that is important for the further use of the digital data. There are different possibilities to deal with this problem. In the first example some kind of natural breaks may be used to define the borders of a region object. Such cases are rare in the real world. As far as for example the landscape is concerned, only manmade changes apply to this case. Streets or fences can be such objects with sharp edges towards neighbouring entities. The second example shows a continual change represented by interpolated borders. In situations where two or more levels may be differentiated in a more or less clear way, borders can be derived by interpolating the slopes between two levels. Continual changes may also be represented using borders with equal

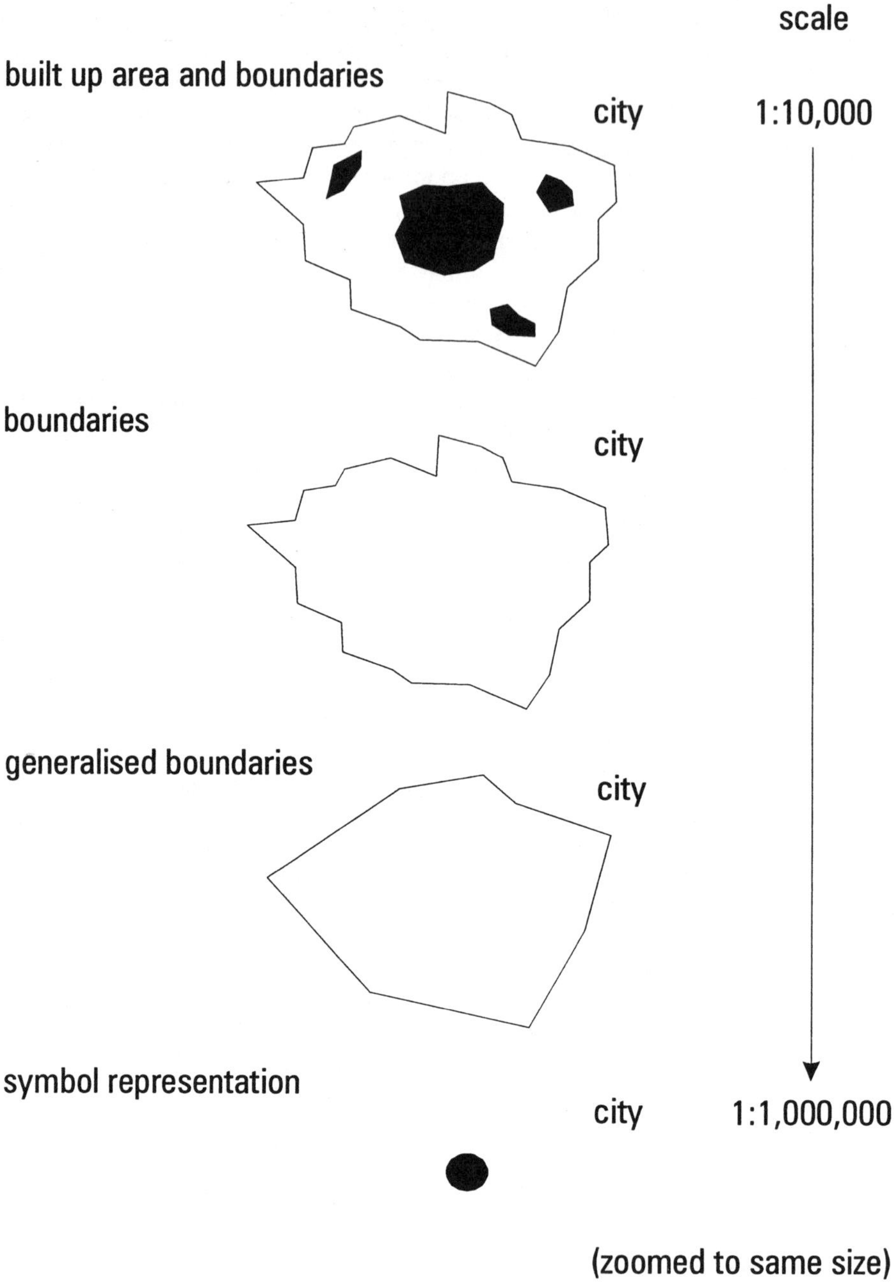

Fig. 4.7 Different levels of scale, related content and accuracy

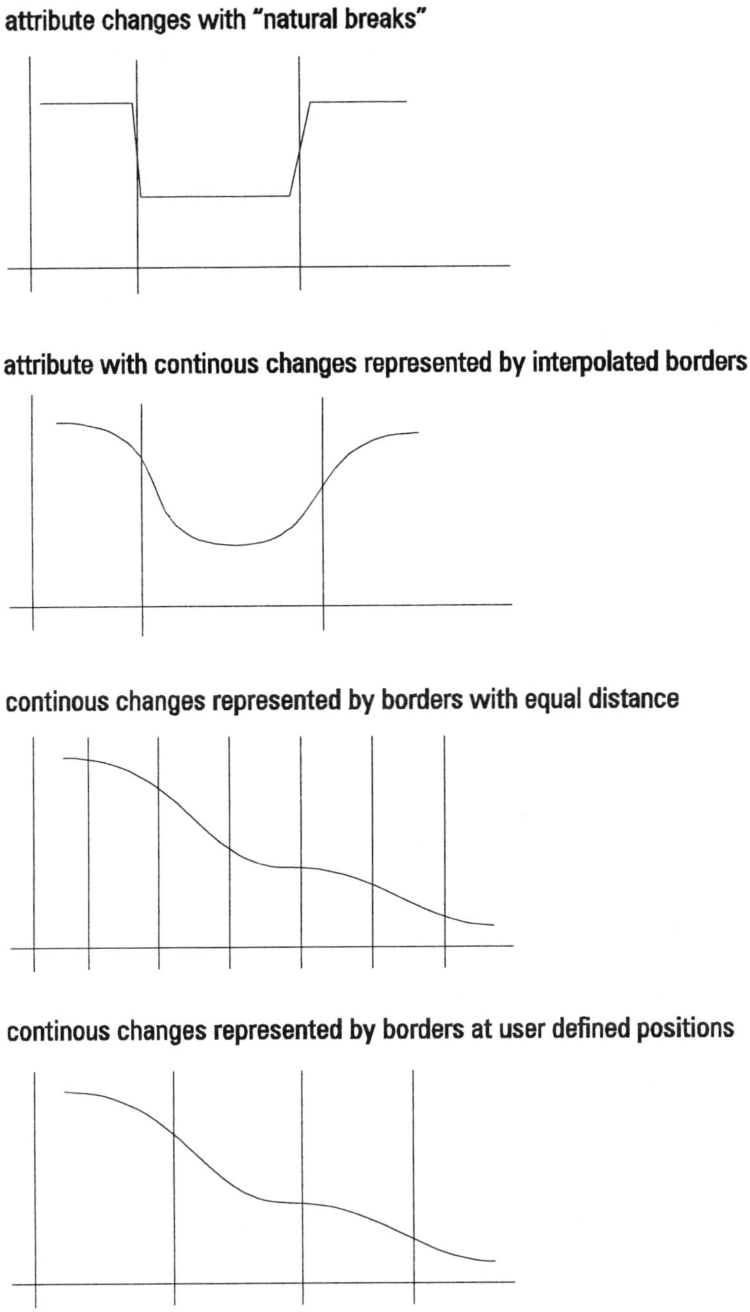

Fig. 4.8 Representation of continuous changes by discrete borders

distances between them. In practice this means a lot of often unused resolution with a high degree of redundancy. In most cases continual changes are represented with borders that use some kind of grouping algorithm. A slope could e.g. be divided into regions of a certain elevation as it is applied in the case of contours.

The value of a map or a dataset for a given task depends on how such continual changes are conceptualised. A simple example may describe the importance of this aspect of spatial data. Given is a landscape where a forest continually turns into bushes which are followed by open field, sometimes used as pasture. Depending on the scale it is for most purposes sufficient to interpolate the border between forest and open field. In our example it would be where the bushes are. There is a scale dependent inaccuracy resulting from the generalization of the forest-open field continuum into only two categories, forest and open field. If we look at the parcels which are used as pasture, we could use the legal boundaries to define the pasture areas. This leads to a second source of error, which evolves from a definition which implies more spatial accuracy than to be found in the real world. The sharp edges of privately owned parcels used as pasture could e.g. in terms of vegetation hardly be separated from neighbouring areas of open field.

This second example shows that definitions which may be useful for one use may not be adequate for another objective. The biologist interested in vegetation will not be satisfied with a coarse differentiation between forest and open field. He will possibly, depending on the scale, introduce several more classes. On the other hand he may not be interested at all in the legal definition of privately owned parcels, as long as there is no special vegetation which could be found there.

The temporal aspects of maps were mentioned before. Here some aspects of temporal data (see chapter 3) have to be discussed according to their importance for the acquisition of spatial data, the difference between the time of an event and the time when this event is included in the dataset. As said before, topographic maps represent only on time step. A topographic map is therefore a generalized snapshot of the real world. Most of the objects included in a topographic map have a more extended lifespan. Some entities of the real world hardly ever change over considerable periods of time. This is for example true for the elevation. Such information is therefore not each time re-evaluated, when a new map is drawn. It is important to know that it takes some time before such changes are included into the new maps and that the date of the last editorial changes of a map does not imply that the whole map is re-evaluated and redrawn. Because such changes are often not otherwise documented, there are two sources of temporal inaccuracy of spatial objects:

- If the change is not documented otherwise, the accuracy is given by the time span, first, between the production of a map before the event occurred, and second, the production of the map after the event occurred.
- The error of the first type has to be added to an error of the second type. This error is defined by the time span between the last re-evaluation and the production of the map, or if this period is not known, the average time between the re-evaluation of a real world entity.

Some thematic maps contain more detailed temporal information. This is the case for example if the occurrence of an event is mapped, like maps of historic battles, archaeological sites etc. Entities may also be grouped temporally, like for instance in geological maps. Such additional temporal information in thematic maps is often mapped without a sufficient level of accuracy. This means that such maps in most cases may not be used to build a GIS. The maps should be treated equal to other temporal attribute data. The temporal information should be stored in a relational database management system and geo-coded with accurate spatial objects gathered e.g. from topographic maps.

Most GIS projects cannot cope with only one source of spatial data. Besides thematic maps which are not accurate enough to be geo-referenced in a GIS there are several types of special maps, often based on topographic maps. As an example, figure 7.4 (see chapter 7) presents a rich variety of maps with different scales and also sometimes different projections and levels of accuracy, which are used for the analysis of cultural landscape changes.

Working with such different sources in one GIS, it is important to guarantee spatio-temporal reliability of the different layers. This means, that identical entities in different source maps or layers have to be represented by only one object in the GIS. After the homogenisation of the different data sources, all identical point, line, and region objects have to be linked to only one object, valid for all layers. Whether such an object may be used once for all layers or has to be stored redundantly depends on the chosen approach and the selected software.

Figure 4.9 gives an example of the application of this technique for one point. The spatio-temporal reliability applies not only to different sources and layers of content, but also to the same object in different scales and at different times steps. In Figure 4.10 the overlay of two region objects is shown. Using a simple overlay technique the inaccuracy of different sources leads to minor inaccuracies. Depending on the scale the differences can clearly be identified and the overlay result is not suitable for mapping. For the user of the map it is not possible to determine, whether the two objects represent two different or one identical object. In the second example the two objects were split in order to perform spatial operations. The result are several sliver polygons. They belong to none of the original objects. Area calculations are now no longer correct, because the sliver polygons are subtracted from the objects where they originally belonged to. Also several spatial operations can now no longer be applied (see chapter 4).

For the necessary homogenisation the map with the largest required scale and the best resolution and accuracy should be the master-layer. The homogenisation can be done in different ways. The least complex case is, when different layers must be matched from different scales but only one spatial division, which means that the maps are only different in scale and therefore accuracy. In this case the most inaccurate layer could be homogenised by converting it into a point layer. While performing this geographic operation the ID number and the related attribute information are assigned to the new objects generated by the GIS. In a second step the point in polygon query could be used to assign this ID and related attribute data to the accurate region layer.

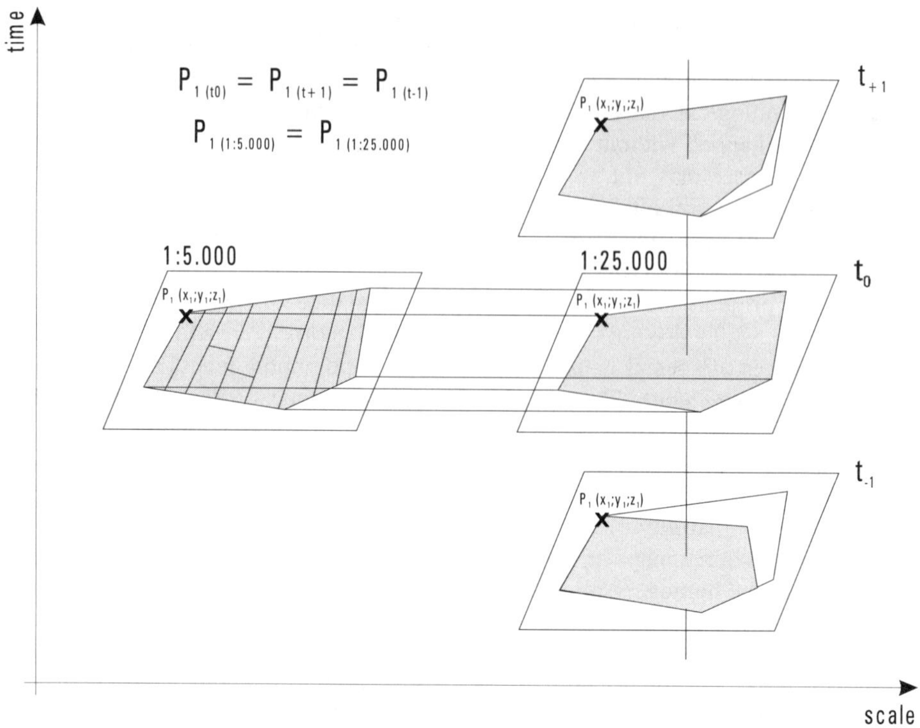

Fig. 4.9 Spatio-temporal reliability

Other cases, where territorial changes happened, request more effort to homogenise the territorial units or other spatial objects. One possible solution is to transform and rescale the inaccurate layer, so that it will match a base map from which the accurate layer is digitised as good as possible. A printout on transparent paper could be used to mount this map upon the base map. The homogenisation starts by copying the identical parts from the accurate layer to a new layer. Missing objects have to be re-digitised. The printout must be moved to fit the existing "holes" in the copied map manually by adjusting the two maps the fact that territorial borders are often bound on linear entities of the landscape is of great importance. Also landscape elements, which have a great durability, could be addressed in this way. It is important that such a manual interpretation of the two maps for the homogenisation are made by an advanced researcher and that additional information is provided to foster the decisions made.

This manual homogenisation of two maps is really time consuming and therefore expensive, without detailed additional information a common source of errors. Older maps without a projection or severe accuracy errors may often only be used together with more modern maps. Sometimes not even an incorrect map is available. Some ancient boundaries are described in words or using sketches. Such maps may be utilized for a GIS, but one has to be very careful using the resulting maps and address the possible errors in the interpretation of the results.

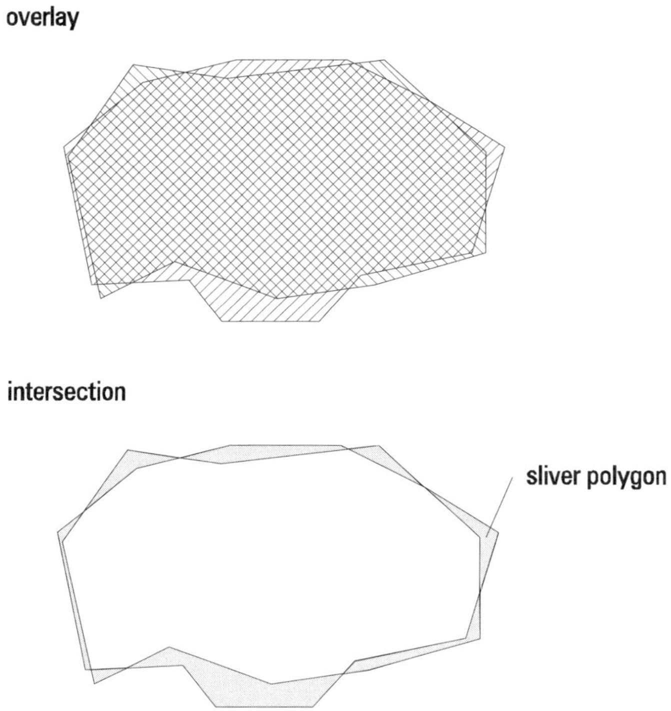

Fig. 4.10 Overlay and intersection of spatio-temporally unreliable objects (adopted from DeMers 1997:343)

4.3.1.3
Territorial and areal unit changes

The majority of spatio-temporal data from censuses, surveys et cetera is in a spatial respect related to some kind of areal or administrative unit, often represented by a region object in a GIS. Each object in a GIS has a temporal validity which should be defined in a related database or meta-database. The temporal validity may be defined by exact time stamps for the birth and death of the object. Either birth or death may also be undetermined, open, or not exactly defined. The matrix in figure 4.11 shows some of the possible combinations. The validity of an object defined by birth and death for a given time step may be derived easily using Boolean queries (see chapter 5.2).

The modelling of continual changes of territorial or areal units with an Entity-Relationship-model was described in chapter 3.2. This chapter deals with the sources of information to detect such changes. In historic times such changes of territorial units occurred in most countries. Several countries have a long record of such changes until today. In Germany for example there is a continuous change in county boundaries since the World War II. In most of the federal states, founded

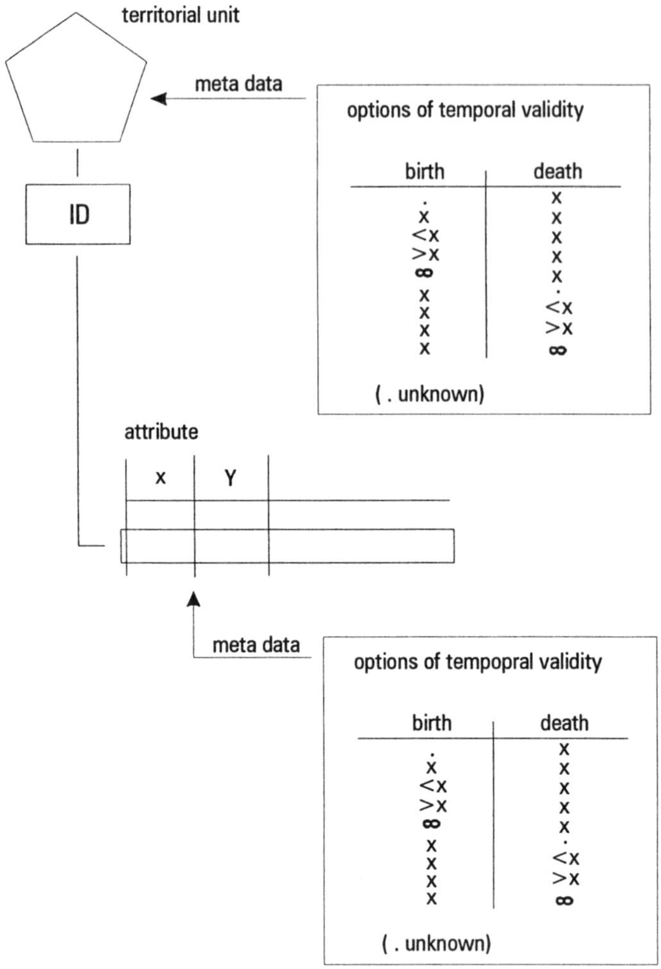

Fig. 4.11 Geo-coding of temporally determined objects

after the establishment of the Federal Republic of Germany in 1949, there was a major restructuring of the municipalities in the 1970s (see figure 4.12 and 4.13), which sometimes also affected the boundaries of the counties, the territorial units on which most of the variables from the statistical offices are based on. With re-unification the districts of the GDR ceased to exist and the former federal states were reintroduced. The counties were afterwards subject to numerous changes, sometimes affecting the boundaries of the federal states. The practical implications of such territorial changes are obvious. The corresponding statistical data can no longer be transformed into time series because of the changing areal units the variables are based on.

In modern times there is always an exact temporal definition of each territorial change. The change is usually stored in the responsible legal record office. In Germany the responsibility for such changes belongs to the federal states. Besides the legal act of altering the boundary of a municipality, city or county, or the merging of two or more communities into one new community additional information is stored (e.g. population affected by the change, the exact area which was exchanged between the territorial units, the new ID number which is used for the official statistics) (see figure 4.14). Such complete data covering each event of territorial changes may be stored in an E-R-model and used to generate the valid state of territorial units for every possible time step without any problems (see figure 3.12). Using such detailed information it is no problem to transform statistical data for different time slices with a different territorial state into time series.

Unfortunately the collection and processing of the territorial changes from the legal records or information published by statistical or survey offices is time consuming and expensive. Sometimes such data is already available for purchase but it is in most cases too expensive for small projects. As a rule, there is no information on the boundary changes in such records of territorial change. As long as the boundaries of the new territorial units can be generated by recombining older territorial units, the update of the geometries is no major problem. Selecting old geometries and merging to a new geometry with new ID number can easily be performed using the macro languages of any standard GIS. If there is a complete change of the boundary, new geometries have to be digitised from appropriate maps or acquired from data vendors or the survey office.

Another source of information on spatial changes are gazetteers which are published yearly or at the occasion of a census. Gazetteers describe all communities or other top level administrative units at a specific time. However, compared to the records of territorial change the gazetteers do not account for the changes that occurred between two editions of a gazetteer. The changes have to be derived from the differences between two gazetteers. Because there is usually no information on the spatial relationship between the old and the new units, time series can not be compiled from such data. Changes can only be described by attributes like “new unit”, “unit still existing”, “ceased unit”. Such gazetteers typically include small scale administrative maps, which may be used for digitising if accuracy and scale are sufficient for the task.

A more detailed spatial resolution is provided by topographic maps. Besides the topographic information, topographic maps usually display the administrative boundaries at their exact position. However, a disadvantage of topographic maps for the detection of territorial unit changes is, that topographic maps are only updated from time to time. Usually a new edition is prepared when the content of the map has changed considerably. In general, territorial changes are followed by new map editions. Unfortunately the update is not always made immediately after the changes took place. So the only precise information which may be taken from a topographic map is, that a particular border is temporally valid for the time the map was made or published. Some modern topographic maps have additional information on the temporal validity of territorial units shown.

Fig. 4.12 Administrative boundaries of the federal state of Baden-Württemberg before 1973

Fig. 4.13 Administrative boundaries of the federal state of Baden-Württemberg after 1973

	01. 01. 1974	Neuer Gemeinde-schlüssel	Neue Gemeinde	Neuer Kreis
	SAARLAND 10	100		
	Kreisfreie Stadt SAARBRÜCKEN Landeshauptstadt 10 011 000	10 041 100	Saarbrücken, Landeshauptstadt	Stadtverb. Saarbr.
	LANDKREIS HOMBURG 10 0 31	10 0 45		Saar-Pfalz-Kreis
10 0 31	111 Bexbach, Stadt	10 045 111	Bexbach, Stadt	Saar-Pfalz-Kreis
	112 Böckweiler	45 112	Blieskastel, Stadt	Saar-Pfalz-Kreis
	113 Brenschelbach	45 112	Blieskastel, Stadt	Saar-Pfalz-Kreis
	114 Einöd	45 114	Homburg, Stadt	Saar-Pfalz-Kreis
	115 Frankenholz	45 111	Bexbach, Stadt	Saar-Pfalz-Kreis
	116 Höchen	45 111	Bexbach, Stadt	Saar-Pfalz-Kreis
	117 Homburg, Kreisstadt	45 114	Homburg, Stadt	Saar-Pfalz-Kreis
	118 Jägersburg	45 114	Homburg, Stadt	Saar-Pfalz-Kreis
	119 Kirkel-Neuhäusel	45 115	Kirkel	Saar-Pfalz-Kreis
	120 Kirrberg	45 114	Homburg, Stadt	Saar-Pfalz-Kreis
	121 Mimbach	45 112	Blieskastel, Stadt	Saar-Pfalz-Kreis
	122 Oberbexbach	45 111	Bexbach, Stadt	Saar-Pfalz-Kreis
	123 Webenheim	45 112	Blieskastel, Stadt	Saar-Pfalz-Kreis
	200 Amt Altheim			
	211 Altheim	45 112	Blieskastel, Stadt	Saar-Pfalz-Kreis
	212 Neualtheim	45 112	Blieskastel, Stadt	Saar-Pfalz-Kreis
	300 Amt Breitfurt			
	311 Bliesdalheim	45 113	Gersheim	Saar-Pfalz-Kreis
	312 Breitfurt	45 112	Blieskastel, Stadt	Saar-Pfalz-Kreis
	400 Amt Limbach			
	411 Altstadt	45 115	Kirkel	Saar-Pfalz-Kreis
	412 Kleinottweiler	45 111	Bexbach, Stadt	Saar-Pfalz-Kreis
	413 Limbach bei Homburg	45 115	Kirkel	Saar-Pfalz-Kreis
	414 Niederbexbach	45 111	Bexbach, Stadt	Saar-Pfalz-Kreis
	500 Amt Medelsheim			
	511 Medelsheim	45 113	Gersheim	Saar-Pfalz-Kreis
	512 Peppenkum	45 113	Gersheim	Saar-Pfalz-Kreis
	513 Seyweiler	45 113	Gersheim	Saar-Pfalz-Kreis
	514 Utweiler	45 113	Gersheim	Saar-Pfalz-Kreis
	600 Amt Walsheim			
	611 Niedergailbach	45 113	Gersheim	Saar-Pfalz-Kreis
10 0 31	612 Walsheim	45 113	Gersheim	Saar-Pfalz-Kreis

Fig. 4.14 Record of changes of the federal state of Saarland 1974

4.4
Spatio-temporal GIS approaches

The following section aims to facilitate the implementation of spatio-temporal GIS under constraint conditions. The approaches presented are related to practical needs and should encourage the use of GIS for the analysis and visualisation of spatio-temporal data.

All examples are chosen for the use with standard desktop GIS packages and other common office software. In the first instance, there is no need for considerable investments to profit from the examples given. Additional value may be added to existing data by simply applying a new spatio-temporal view on the data. Some visualisation may even be done without the need of a GIS, using spreadsheet and drawing software. The complexity of the projects presented here for the utilization of given datasets can be adjusted to specific needs and financial and personal resources available.

The first example deals with temporal point maps. This is one of the most basic concepts to visualize temporal data and can be applied in all cases where temporal information is part of attribute data. Because most data posses a spatial representation such attribute information can be mapped by using simple point-objects or spatial coordinates to geo-reference the data.

In the second example gazetteers are used to build a framework to use historical statistics or census data as spatio-temporal data. The gazetteer method can be applied if there is no exact basic spatial information which could be used to geo-reference available data. It is often not only necessary to deal with insecure spatio-temporal relations but also to build time series from statistics of different origin or censuses from different years. For this purpose common attribute information like place names, ID or postal codes which are not constant over time, have to be homogenised and made available to geo-code the respective data via the appropriate attribute information with a spatial expression.

A completely different approach has to be used for a convenient modelling of spatio-temporal processes. Raster cell information can be converted from vector layers, models can be build e.g. by using point surfaces or Thiessen polygons derived from points. Mathematical models may be applied to such target layers to calculate spatio-temporal projections. Finally a grid cell overlay is the easiest method to track temporal changes over time.

In the next section snapshot maps with temporal content are targeted. Most maps contain spatial objects with temporal information. For example the age of a city, railway line, street etc. in a topographic map. The temporal information in such snapshot maps may not be given explicitly. Sometimes it is possible to collect this information from external sources like archival records. Another source of such information are consecutive layers of topographic maps, from which the live span of a spatial object may be determined. This past in present maps are important for all fields of application where objects in the real world, coming down to us from the past.

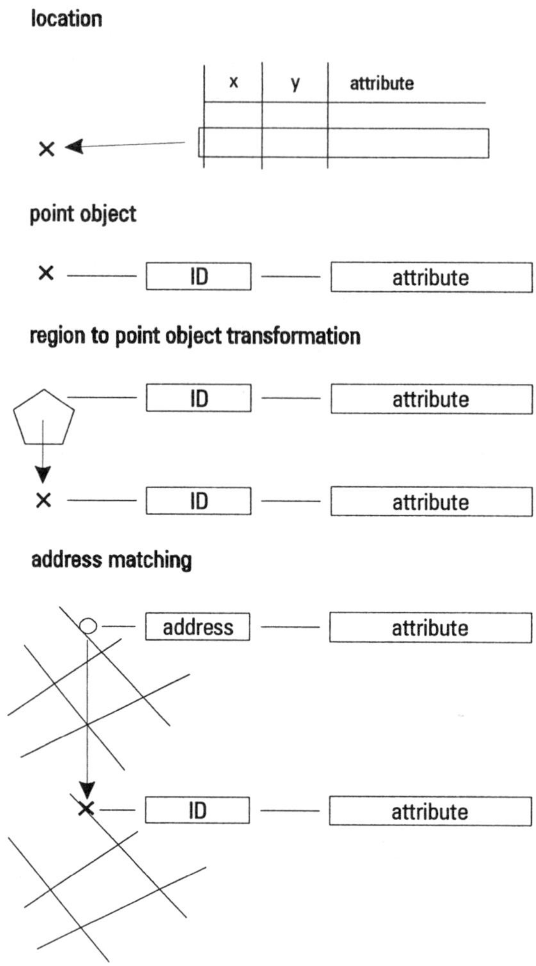

Fig. 4.15 Spatial reference and attribute sources for point maps

A common GIS analysis method, layer overlay combined with the intersection of vector objects, is utilized in the next approach to convert mono- or multi-temporal data. In traditional mono-temporal GIS such area weighted splitting techniques are used for instance to homogenise data based on different spatial resolutions.

The following method presents a more sophisticated approach. In chapters 2 and 3 numerous requirements were stated, which justify the emphatic demand for the application of an object-oriented GIS for the conceptualisation of spatio-temporal processes and the analysis of the spatio-temporal data. Only a system that manages the spatial data completely in an object-oriented fashion on the logical level, is able to fulfil the necessary requirements. It must be temporally and thematically flexible and adaptable it must be scale-independent and it must be able to be extended at a later date. Unfortunately such systems are not available in the desktop GIS market at present. Therefore the conversion of the conceptual data model on the logical level should be accomplished via the combination of a layer-oriented GIS and a relational database management system.

While the previous approach may be applied with each GIS using region objects stored in layers, the following attempt to utilize a spatio-temporal entity-relationship-model as a logical data structure is only possible with packages like ArcInfo, which are able to work with single line segments (arcs). In this case the entity-relationship-model can be emulated in a relational database management system and linked to a GIS using a more or less extended set of software tools to facilitate the creation of layers for each possible time step. Compared to the approach of least common geometries before, which uses a consecutive set of time slices to model a continual period of time, the pseudo-object approach can directly address the spatial changes occurring during a continual period of time. Both approaches are best suited to deal with region objects like territorial area units or administrative boundaries. The advantage of the latter is, however, that changes and updates are easier to maintain.

Fully object oriented spatio-temporal GIS are complex and may not yet be applied in ordinary GIS projects.

4.4.1 Temporal point maps

Point data are among the most frequently used spatio-temporal data (see figure 4.15). They may be provided in different forms and formats, such as:

- A list of points with geographical coordinates.
  In the most simple case x- and y-coordinates as well as associated attribute data are stored in a database table and later processed in a GIS. Geographical coordinates result for example from surveying. If a GPS receiver is used they may be directly imported into a GIS.
- Geo-coded point-objects obtained from manual digitisation.
  Using an identifying feature, attribute information can be linked easily to geo-coded point objects obtained from manual digitisation.
- Geo-coded point objects transformed from region-objects.
  Attribute data connected with region-objects may be transformed easily into point-objects. Depending on the algorithms used by the GIS, the location of the resulting centroids may vary.
- Geo-coded point objects derived from address-matching.
  Using a geographically referenced road network inside the GIS, attribute data which is only pinpointed by a street address may be assigned to point objects by address-matching procedures.

Based on these point objects, different kinds of spatio-temporal attribute data may be visualized and processed in the GIS (see figure 4.16). The temporal aspects may be classified as follows:

temporal content	stored attribute data
discrete linear time event	t = 05/04/2000;1999; ...
discrete circular time event	t = "monday"; "spring"; "August"; 23:55 ...
event classified by period	t = "holocen"; "iron age"; 1991-1999; ...
changes during period	$t = f(t_1; t_2)$
time series data	$t_1 = x_1, t_2 = x_2, t_3 = x_3, \ldots, t_n = x_n$

Fig. 4.16 Example of temporal attribute data for a point maps

- Points in time when an event occurred (e.g. date, year).
  Events may be either characterized discretely by calendrical time or assigned to temporal periods, which may even be marked off in a fuzzy manner.
- Changes between two points in time (expressed e.g. in proportional values).
  Information on the variation between two points in time may be derived from any variable which is available for at least two points in time.
- Temporal sequences.
  Temporal sequences can be described by statistical parameters or visualized in a graph.

Spatial point maps represent the most simple form of visualizing spatio-temporal data. In this case the visualization functions of the GIS are predominant. The production of temporal point maps makes little demands on hard- and software, as well as the quality of the data to be displayed. They can also be generated with drawing software or digital mapping software. The derivation of statistical measures in order to characterize changes or to describe temporal sequences is made with the help of standardized functions which are also available in regular spreadsheets or statistical software packages. The following paragraphs discuss some examples and applications of these maps.

Figure 4.17 depicts the location of an object at different points in time, and thus the spatial variation of this object in a given time interval. Maps showing the shifting mean centre of a population (e.g. for different census years) are a widespread example of this technique. This form of temporal point maps is also suited to visualize any kind of spatial motions and shifts. Using the analytical functions of the GIS, the user may calculate spatial parameters like the distance covered in a certain period of time.

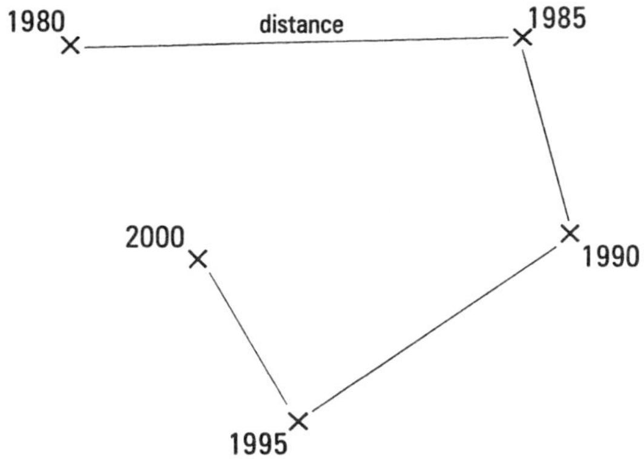

Fig. 4.17 Point maps: single object – location at different time steps

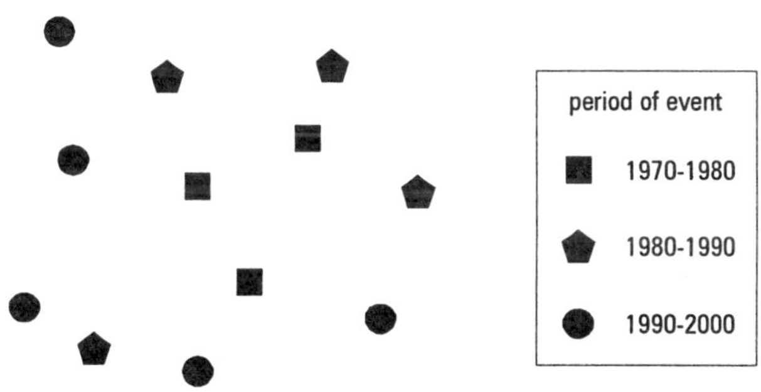

Fig. 4.18 Point maps: multiple objects – location of events grouped by periods

Figure 4.18 displays the location of several objects. The occurrence of events during different periods of time is mapped by symbols. This method of visualization is particularly suitable for spatial diffusion processes. The different stages of diffusion can be interpreted easily from the different symbols used. Again, distances can be measured and spatial parameters may be calculated and analysed. By classifying and grouping of the point objects it is possible to assign both, non-calendrical information and periods of time that are not clearly marked off. The dating of geologic formations or the allocation of archaeological artefacts may serve as an example.

Figure 4.19 portrays another example of mapping temporal changes: Increase and decrease are represented by diverging shadings, the changes in percent are embodied by different symbols. The symbol size depicts the values of a third variable at a certain point in time, for example the base year of the time series displayed. Such maps are frequently used to document changes, e.g. of the number of inhabitants in a region. Again, various analyses may be performed using the functions provided by the GIS.

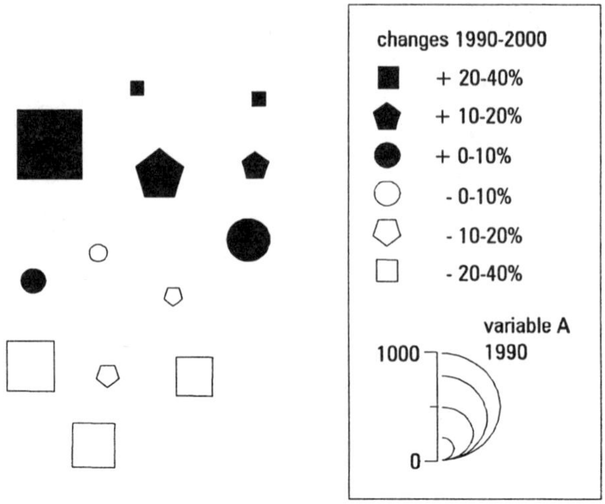

Fig. 4.19 Point maps: single object – location at different time steps

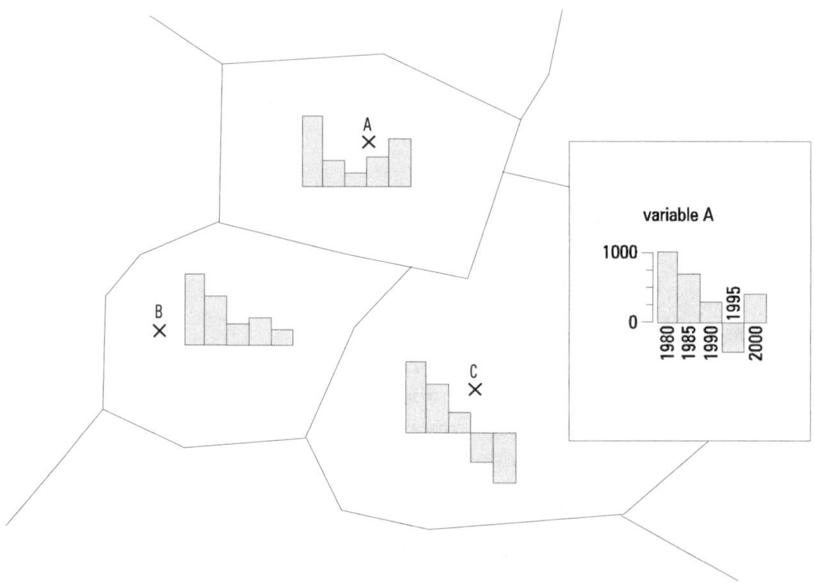

Fig. 4.20 Point maps: single object – time series by location

Finally, figure 4.20 displays time series data as a cartogramm. The diagrams are located geographically exact, a fact which makes the interpretation of the map easier. Such maps may be produced using very basic techniques, e.g. by using a scanned base map as a background.

Temporal point maps make only little demands on the underlying spatial reference systems, e.g. in terms of accuracy. As a consequence they may be used to locate either data with an imprecise spatial definition or data which is related to spatial objects with inexact borders. This holds especially true for historical maps and data, where the spatial reference system may not be restored at all or only by investing unreasonable efforts. Temporal point maps may be used to visualize almost any kind of data that is not related to an area. However, the possibilities to represent spatially inaccurate objects by points are depending on scale. There is no clear dividing line between a geographically exact map and a simple diagram.

Another field of application for temporal point maps results from the principles of map generalization, i.e. areal objects that may not be visualized correctly on small scale maps. Census maps that display the number of inhabitants of the municipalities by using shaded proportional circles are a common example. Other potential uses emerge, if more complex techniques are not necessary or impeded by the specific characteristics of the data. Temporal points maps may then be used to reduce costs.

In many cases information associated with areal units may be assigned to point objects. This may be due to the generalization problem, as the visualization of attributes a choropleth map may not be possible in a small scale. Often the attribute data is not connected with an areal unit; thus, the construction of a choropleth map would be incorrect. In this case, centroids may be derived from the areal units and associated with the attribute data.

The techniques described above are common features of most commercial GIS packages.

4.4.2 Gazetteer Method

For the vast majority of historical data a spatial reference is only provided by a place name. As place names often change over time, it is in many cases impossible to build time series of related attribute data (see figure 4.21). Furthermore, the spatial reference is frequently altered by administrative or territorial changes such as incorporations or fusions of administrative bodies. Often such changes are not documented properly; a reconstruction from archival sources is usually time consuming and expensive.

Information on the respective status of territorial units is usually recorded in gazetteers. These documents list for example all municipalities of a territory at a cut-off date, often supplemented with additional information on area, population or administrative affiliation. In modern times the territorial units are identified by a more or less complex system of key numbers, which may serve as a link to statistical data collected by governmental agencies. Each census automatically embodies a gazetteer, as it contains a complete list of the territorial units included.

Point objects representing the spatial entities covered by the census can be derived from associated census maps or by transferring the information onto modern maps (see chapter 4.3.1). Attribute data associated with the spatial entities can be assigned to the point objects by geo-coding the names.

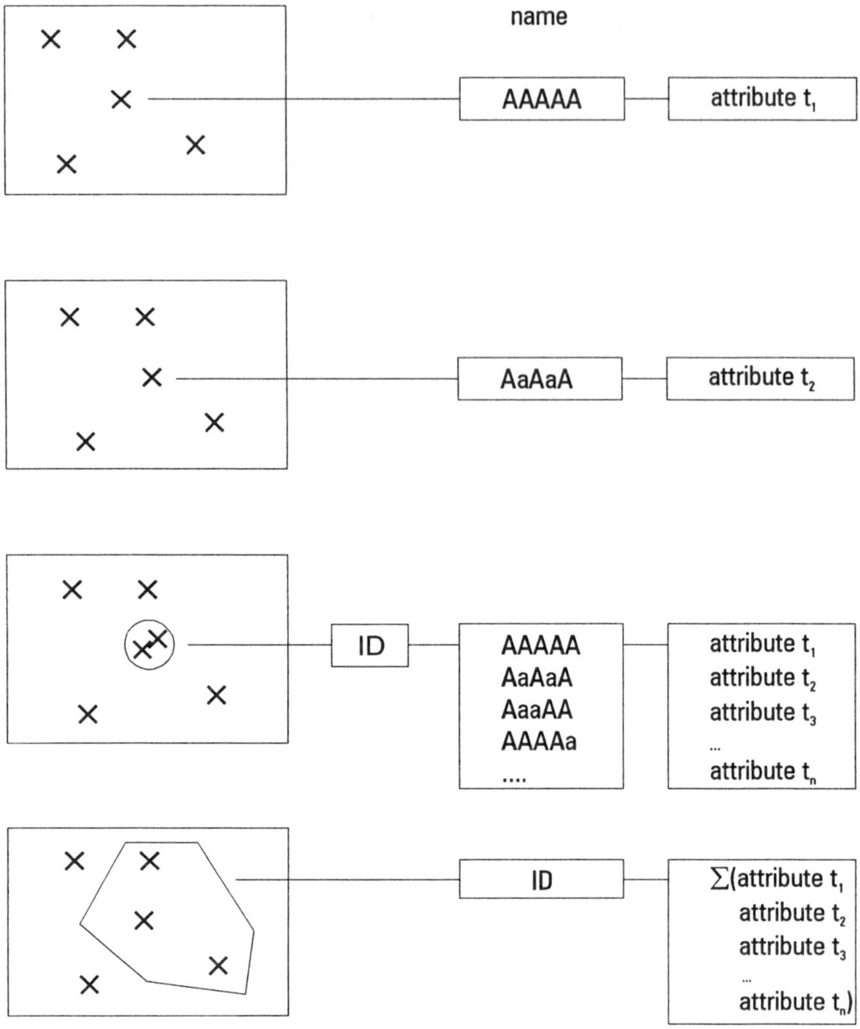

Fig. 4.21 Gazetteer method

In general, identical spatial objects are frequently collected redundantly, either due to the use of maps of different scales, projections and accuracies, due to inaccuracies of the digitisation process, or due to changed spelling of place names. The consolidation of the data holdings can be made with the help of spatial relations: objects with a neighbouring distance inside a certain range of tolerance may be filtered using the standard functions of a GIS (see chapter 5.2). Following a manual plausibility check the differing point objects may be aggregated into a single object, taken from the most precise source map. The subsequent geo-coding process is implemented in two stages: First, the ID number of the final point object is associated with a database table of valid variations (e.g. name spellings). In a second step, the attribute data of the variations is assigned to the final point object by using this translation table. This process is, however, subject to the limitations that apply for point objects in general (see chapters 2.2 and 4.3.1).

The gazetteer-method is frequently used with the aim to explore historical developments using time series of statistical or census data. However, geo-coding the related attribute information is only part of this process. In order to associate the geo-coded attribute data to identical spatial units, the territorial changes that took place in the course of history must be reflected. Very often, such time series need to be continued from historical dates to the present. Thus, the historical attribute data associated with historical territorial statuses need to be applied to a consolidated modern territorial status. With some often distressing exceptions to the rule, this task is facilitated by the fact that present administrative units may be composed by aggregating historical spatial units. Using the latest territorial status as a region object the point objects inside this region may be aggregated together with the associated attribute data in order to obtain time slices of historical attribute data related to the present territorial units. Depending on scale and accuracy of the maps used, it is however very important to perform manual plausibility checks, in order to prevent wrong assignations.

Using parameters of location the gazetteer-method saves a lot of work with the construction of time series from historical data holdings that reflect territorial changes. Besides the simplified consolidation of data from different sources, the data may be associated to modern spatial entities effortlessly, even if the territorial changes are not documented completely.

The subsequent processing of the spatio-temporal data holdings emerged from the methods described above may be accomplished according to the techniques described for point objects in chapter 4.3.1. In spite of the ensuing limitations, these methods also possess a number of advantages compared to working with region objects derived from a modern map: The operations being applied relate to scale-depending generalizations of historical entities. The consolidation of the data holdings using the method of least common geometries (see chapter 4.4.6) only reflects the re-combination of the spatial objects. Changes in the associated areas are not taken into account. Therefore, attribute values related to areal units must be normalized using the area of the respective object.

4.4.3 Raster cell spatio-temporal GIS

One of the major aims of implementing a spatio-temporal GIS is the ability to perform modelling and spatio-temporal reasoning. These aspects are covered in chapters 2.3 and 5.4. The implementation of modelling and spatio-temporal reasoning in a vector GIS is difficult. The quality of the data is of crucial importance. Especially the quality of historical data is frequently insufficient. As described in the previous chapter it is not easy to geo-code historical data for which only a place name is given.

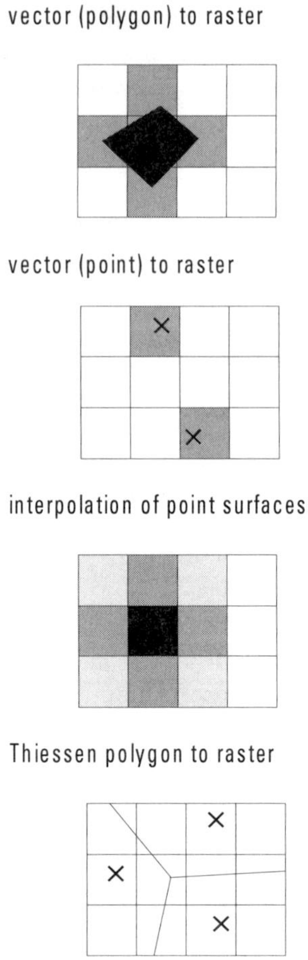

Fig. 4.22 Examples of raster cell conversion

Ancient statistics sometimes cover only major cities, smaller communities are neglected or the records have been lost. In such cases a point surface model may be used instead to try to obtain information for areas between locations for which data is available. Census data for single communities may e.g. be used to derive population densities. On the other hand it is probable that administrative units corresponding with census data are not available. In this case raster cell approaches can be used to calculate area based measures.

If there is no complete vector dataset at hand, modelling and spatio-temporal reasoning with raster cell based GIS is a suitable alternative. In a first step the vector data must be converted to a raster cell matrix (see figure 4.22). Polygon, line and point objects may easily be transformed from vector to raster cells. A lot of GIS packages provide powerful tools for this purpose. The problematic representation of continual developments may be addressed by using point surface interpolations or Thiessen polygons.

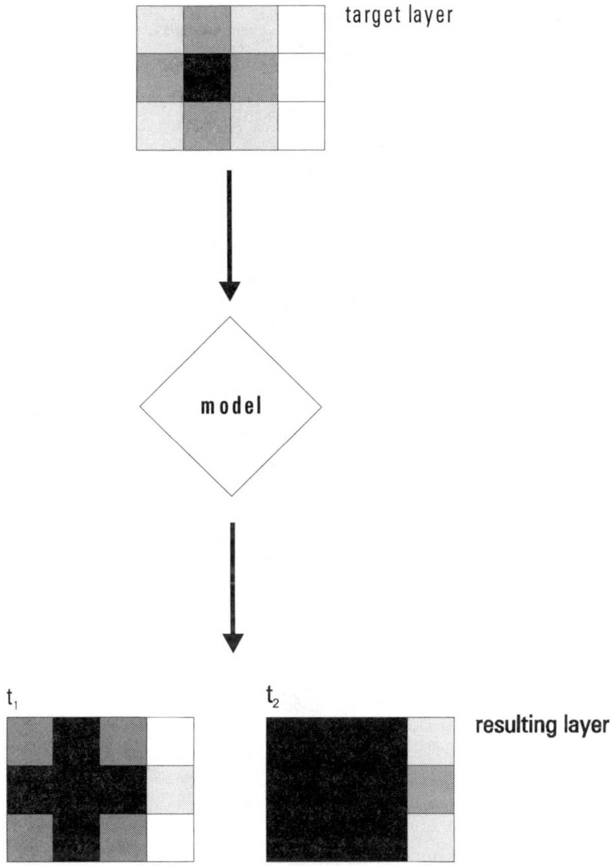

Fig. 4.23 Raster cell based on temporal modelling

The resulting target layer of such actions can be used to apply different kinds of models. Such models may be composed from procedures within the GIS. Often it is more convenient to export the grid cell data to a table which can be processed using external software. For instance a diffusion model can be used for prognostic purposes, generating layers for future time steps (see figure 4.23 and chapter 5.2).

As described in chapter 4.3.1.2 it is often almost impossible to provide identical objects for identical entities in two layers representing two time steps. However, this is a prerequisite for spatio-temporal reasoning in a vector GIS. The detection of temporal changes with a grid cell based system is more versatile and can also be used for data with a lesser degree of spatial accuracy (see figure 4.24). If necessary the grid cell can be generalized conveniently, so that the overlay can be performed. The result of such changes can be computed within the GIS, however, even a spreadsheet program will do the job.

As mentioned above the grid cell approach may be used for all data without a high degree of spatial accuracy and/or without enough resources to process the

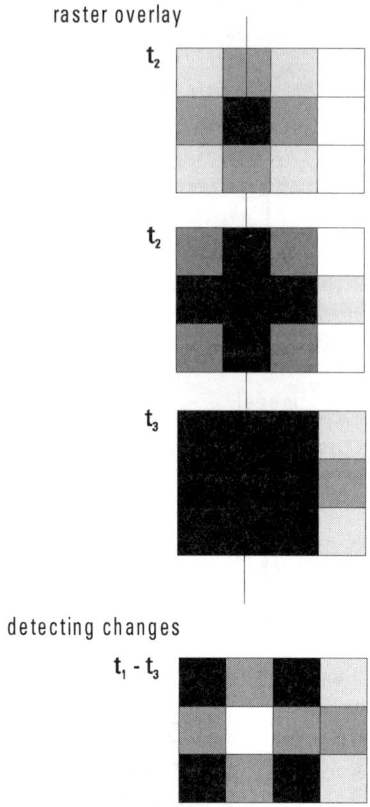

Fig. 4.24 Raster cell based on temporal reasoning

spatial data to fit for a vector overlay. Where temporal changes have to be detected, models applied or spatial interpolation performed, the raster cell method is an adequate solution.

Unfortunately the presentation of the outcome of this method is for most purposes not very efficient. Landmarks such as cities, which are necessary for the interpretation are lost. Furthermore, the original information which was used as input for the grid cell model cannot be mapped as a second information layer. Grid cell overlays using different colours and signatures are difficult to interpret. However, the raster matrix can be split and aggregated to e.g. an administrative layer, in order to include the results in a vector GIS.

4.4.4
Past in present snapshots

Most objects which are part of maps carry some kind of temporal information. Snapshot maps do not only cover the current state of affairs given at the time of

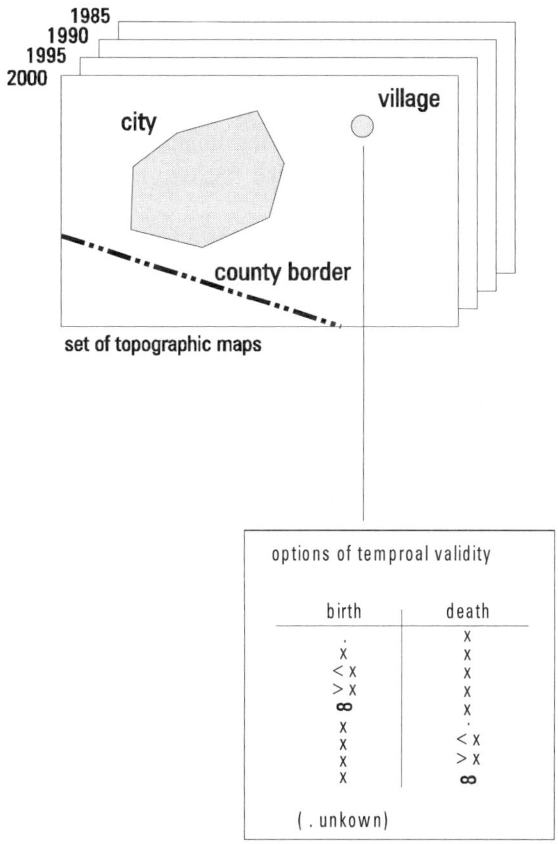

Fig. 4.25 Past in present snap shot

map production, but also a view on the past. This past in present view can be applied to any maps and entities in the real world by asking questions like: "how old are the entities?" or "when did the entities appear for the first time?"

The example given in figure 4.25 displays the symbolized content of a thematic map. A city, a village and a county boundary are shown. In topographic maps the corresponding temporal validity of such objects is in general not given. The age of a city or village has to be acquired by using an archive or other printed material. The birth and the death of the objects can be stored in a table, as shown in the figure. In some cases the birth of an object cannot be determined precisely. Sometimes it is not possible to determine the birth of the object at all, sometimes it is only possible to state, that the object existed before or after a given date or as long as the time span of interest reaches back. As applied for instance in geology, the birth of an object may not be defined by relating it to a discrete time event. Also a grouped period like "Holocene" may be used. For entities of a landscape which are part of topographic maps it is often not possible to get additional information on their age; a second source of temporal information must be used.

A layer of consecutive topographic maps each covering one time slice, may be used to determine the live span of an object. The birth or death of an object can be guessed by comparing the time slices. The termination of the event is given by the period defined by the time slices for which the two maps are valid (see chapter 4.3.1.3).

In chapter 4.3.1.2 the use of thematic maps as a source for geometry data or for digitisation was addressed. The problems of low geographical accuracy is also of importance for the approach presented here. Even more often than in topographic maps, thematic maps or sketches provide temporal information. In most cases, such maps cannot be used directly to geo-reference and geo-code the desired information. They can only be used to derive the temporal data which lacks in topographic maps.

One problem evolving from the changes taking place during time is, that objects between two time slices may not be stable. A city for instance grows; the city which is part of a map for 1990 will not be the same city which is part of the map for the year 2000. Whether such developments can be treated using only one object, depends on the conceptualisation of the real world. If for example the growth of the city is not of interest, the development can be neglected and the spatial representation be aggregated.

The past in present view on maps and real world entities are of high importance for all fields of application where objects in the real world are under investigation, coming down to us from the past. Another common application is the aim to determine the age of objects, for which no additional temporal information is available and where the age has to be derived from their appearing on maps.

4.4.5 Time slice overlay method

From time to time administrative changes occur, with the result of a break of the time series of statistical data. It is often desired to convert the time series in a way to cover the complete period irrespective of boundary changes. For this task two layers representing administrative divisions before and after a change are needed. The spatial data has to be spatially and temporally reliable (see chapter 4.3). Both layers are overlayed and the region objects are intersected; a task that can be performed with any GIS (see figure 4.26). Each spatial change results in at least one new object, so that, in the end, a layer exists from which both layers might be rebuild by simply merging the region objects (see also chapter 5.2).

In a second step the attribute values are re-assigned to the new objects using an areal weighted disaggregating method. The attribute values can be recalculated according to the respective new territorial division. The function is implemented in most GIS packages and in mono-temporal GIS (e.g. used to aggregate and reassign data based on different spatial resolutions).

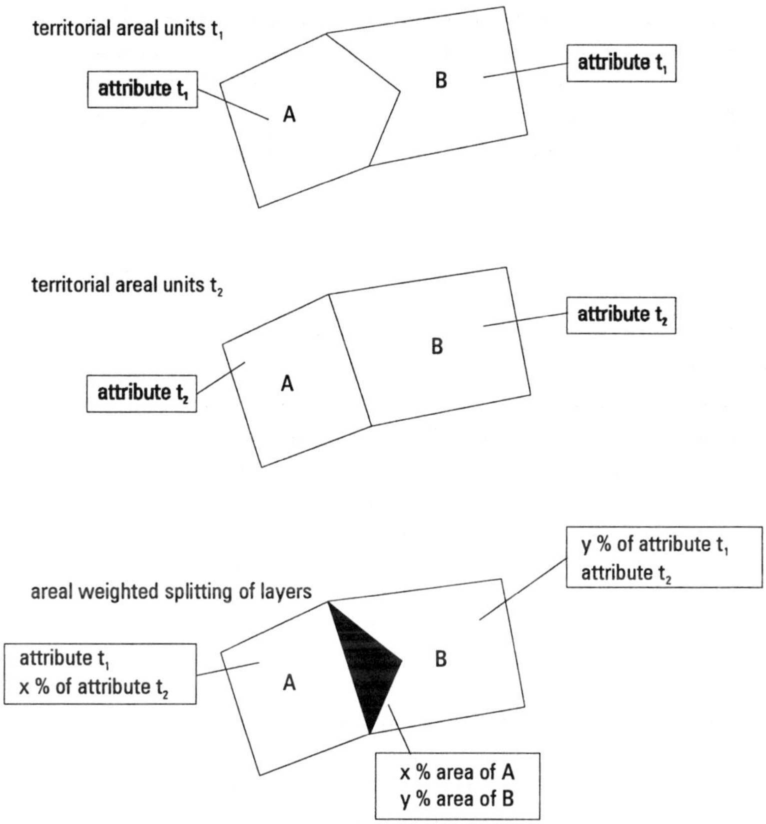

Fig. 4.26 Time slices overlay method

The areal weighted disaggregation of time slice overlays is a convenient way to deal with the problems described above. Nevertheless, there is one assumption made, which is not true in most empirical situations. The dispersion of the value which is disaggregated must be evenly distributed over the whole area. However, most territorial units used for statistics are far from being homogeneous. There is no even distribution of phenomena over space. The error resulting from this false assumption is scale dependent and influenced by the concentration or central-peripheral gradient of a given distribution.

A city with its surrounding area may serve as an example. Given the administrative area of a city, the area consists of the city itself within the border of the built up area. From the centre to the outskirts we may observe a central-peripheral gradient of population density, which varies from city to city but also from region to region and from time to time, as may be learned from urban geography textbooks. Outside the boundaries of the built up areas, still belonging to the administrative area of the city, a sparsely populated area may be identified. The population density gradient may however be broken by suburban residential districts. The example shows, that the assumption of even distribution over space is hardly ever correct. The error can be minimised, if it is possible to use additional information to cover the uneven distribution, e.g. a map with the functional differentiation of the city. A model approach may be applied, relocating the complete population, according to the functionally differentiated areas of the city to smaller more homogeneous areas (see chapter 2.4).

The technical use for this approach is multifunctional. It is also used in the least common geometry method presented in the next chapter. As high quality data is required, the utilization of this technique is most useful with current administrative changes and reasonably accurate maps. Then the effort to prepare the spatial data is not too expensive and time consuming.

The need to compare values for territorial divisions before and after a administrative change is omnipresent, also the completion of time series data. The additional advantage compared to the gazetteer method is that the aggregation of data and the completion of time series data may be used based on the most aggregated territorial division. Furthermore, the time slice overlay method allows to map new data based on older territorial divisions: time series based on ancient administrative units may be continued.

4.4.6
Least common geometries method

A basic problem with the logical representation of spatial objects in a vector GIS is the spatial and temporal reliability of the data model. None of the previously presented approaches addresses this problem in an adequate way. At best they provide an easy work around method to process existing data with the least possible effort and costs. All approaches have considerable disadvantages, one is that they are not able to cover occurring changes continuously. The least common geometry method enables the user to build a GIS based on a quasi continuous space-time composite (see chapter 3.2), which:

- is spatially and temporally reliable,
- covers not only time slices but also continual records of changes,
- can generate boundaries for each given time step, and
- can be implemented with each layer GIS.

With an object-oriented logical data model this goal can be achieved by building all spatial objects from single points. Since individual objects are defined exclusively via their relation to single points, different objects can access identical points without problems. Furthermore, as automated algorithms allow the reduction of redundancy and the concentration of identical points into the most accurate source, the problems of data acquisition from different sources with divergent spatial accuracy can be ignored.

However, the data storage and management in a layer-oriented GIS differs completely as it is necessary to keep identical points, lines, regions or sections in each thematic or temporal layer (often in a redundant fashion). Therefore the object-oriented logical data model must be transformed and split into a discrete layer model. Spatial and temporal reliability may be achieved by the application of the principle of least common geometry (LCG): All spatial objects of a layer model are overlayed and intersected, until we receive items (LCG) that can be reassembled into all elements of all layers (Kilchenmann 1991:14-15; see figure 4.27). However, extending the resulting set of data with further thematic or temporal levels is virtually impossible. Thus, planning must be very far-sighted.

The generation of least common geometries and the subsequent recombination to layers can be elucidated by an example of administrative boundaries (see figure 4.28). It also applies similarly to any other thematic or temporal levels of a layer GIS. In order to produce temporal layers, which do not necessarily correspond with the available time sections of topographic maps, the development of administrative units must be reconstructed from both, available maps (time sections) and other secondary sources.

In the first instance the modifications of the administrative units are entered into a database in chronological order. Both the type of modification and the units involved must be entered in such a way that each modification of a boundary or the area, the date of modification and the involved administrative units can be identified. The database should be structured in a way that for each administrative unit both beginning and end, as well as time and extent of each modification is accessible. The area of the modification serves as a checksum during the entry of the spatial information. It is not only necessary to store these spatial modifications but useful to store the key numbers, which are used to link the thematic data as well.

Second, the current status of the administrative units is entered by vector digitisation based on a map with the largest available scale. Attention must be paid to a most accurate digitisation.

Third, for each modification a new layer is created. In this layer the modifications are digitised from suitable map material in a way that identical lines are duplicated from the data stock already entered. Distortions, projection errors or other inaccuracies must be corrected visually or transferred manually to the reference map material in order to be digitised there.

Fourth, the LCGs are generated by overlaying and intersecting these layers. At least one additional new object results for each modification. A new key number is assigned to each new object. In the database the original object (to be reassembled from the new LCGs) is stored by a reference to the new key numbers. Modifications, which result from the sole combination of already existing areas and which can be entered without digitisation, are stored in the database correspondingly. As a check, the surface size of the new areas is compared to the surface size of the areas stored in the data base.

Fifth, the resulting layer forms the basis for a progressive digitisation and overlay. The final product is a layer with the LCGs and a database, which contains the information for the recombination of the LCGs to valid administrative units for any historic state.

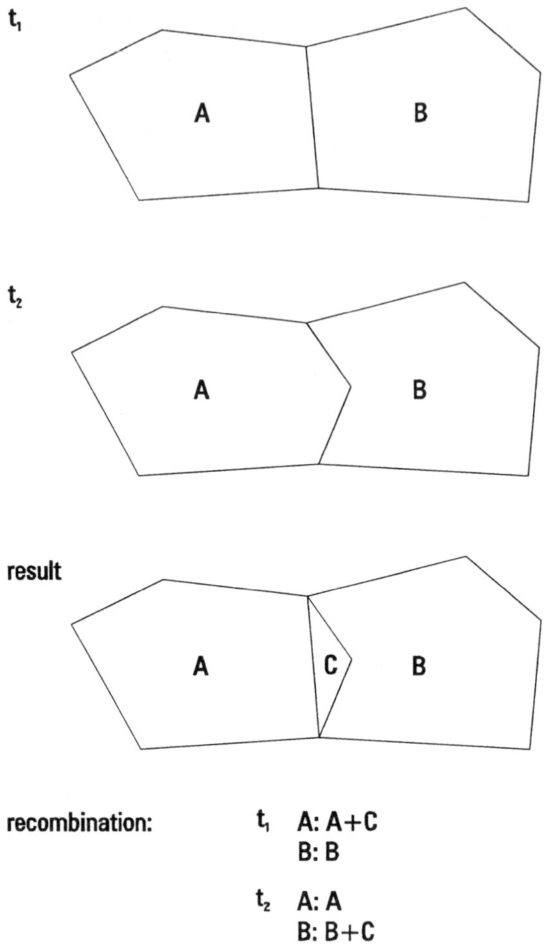

Fig. 4.27 LCG-method: space-time-composite with regions (adopted from Wessels 1997:52)

Sixth and finally, all administrative units, their key numbers and the key numbers of the LCGs which are valid for the respective time section are extracted from the data base to serve as a basis for the recombination of the administrative units from the LCGs. This is executed in a copy of the LCG-layer. Step by step, the LCGs which form an administrative unit are selected, combined and assigned to the key number of the administrative unit.

Two additional points have to be addressed here: the structure of the spatio-temporal database which keeps the information needed to recombine the objects as well as additional metadata including the key numbers used by the statistical offices

The temporal changes are stored in an relational database management system. The structure of this database should be adopted from the entity relationship model of administrative boundary changes, described in chapter 3.2. In some cases

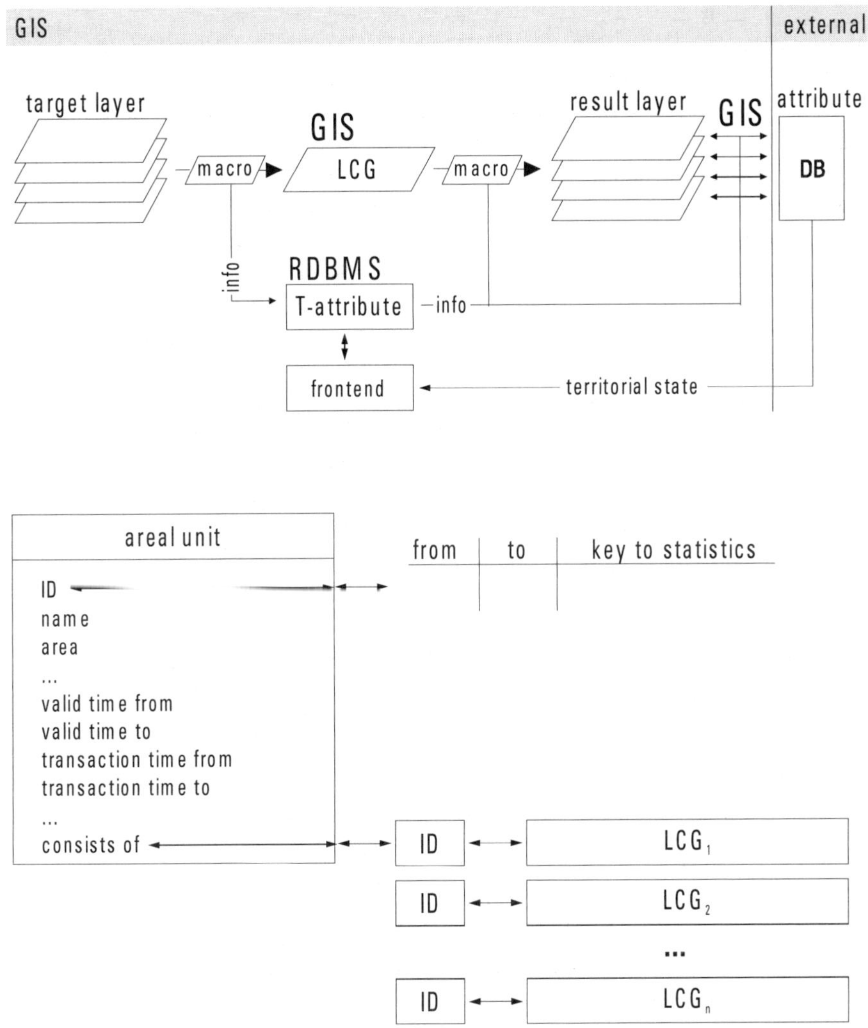

Fig. 4.28 LCG-method: workflow

it is useful to analyse the spatio-temporal changes. Therefore the relationship to predecessors and succeeding objects should be stored, as well as type and reason of the temporal changes. An important information which must be included is the common attribute information used to geo-code the spatial object with the attribute data from external sources. In most cases ID or key numbers are used for this purpose. These ID numbers can change independently from spatial modifications. Different sources can require different ID numbers. The value added to existing data depends on how versatile the GIS may be used to link external data. Therefore a considerable amount of work should be used to provide such interfaces.

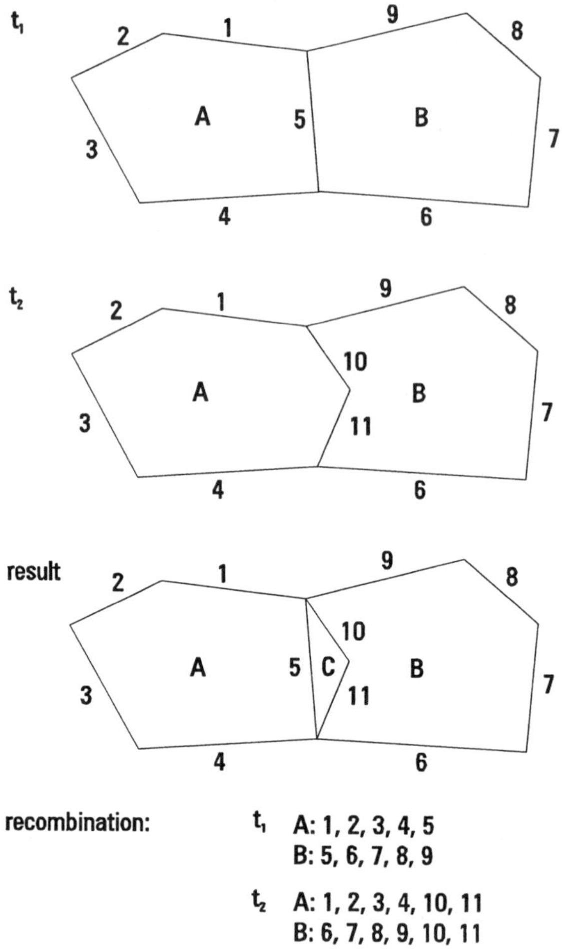

Fig. 4.29 Pseudo-object GIS: space-time-composite with lines (adopted from Wessels 1997:55)

In terms of the temporal information needed to build a spatio-temporal GIS it is necessary that all changes may be clearly identified by assigning a date. Additional precautions must be taken, to include uncertain events. In this case the recombination might lead to an incorrect coverage of spatial units. A location may not be covered by an object at all, the coverage is incomplete, or a location may be covered by more than one object.

The reproduction of temporal sequences in a layer-oriented GIS is extremely complex. If identical boundaries are available, this applies as well to the processing of different thematic layers. The subsequent extension with additional time sections or new thematic layers requires in each case a complete revision of the data holdings. For practical reasons it is almost impossible to add further details in a larger scale.

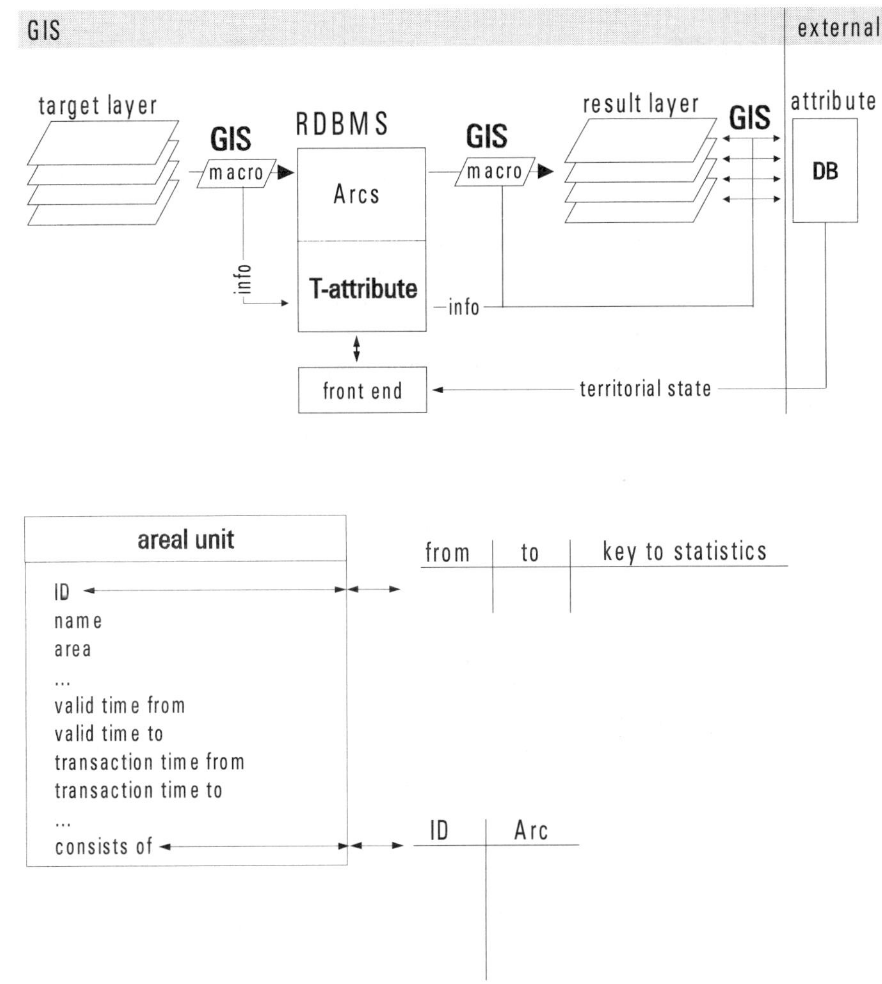

Fig. 4.30 Pseudo-object GIS: workflow

The least common geometry method needs a high degree of spatially and temporally exact data, which is in most cases available only for the 20th century. The processing of the data and the building of the system is time consuming and expensive. Also investments in the development of additional tools to automate the process are necessary if the number of objects or temporal changes is large. Therefore the approach is not suitable for ad hoc analyses. The implementation should not be made in an environment were time and money are not sufficient. However, the advantages of this approach are considerable in terms of versatility and flexibility of analysis. The approach is best suited for projects with the aim to provide services related to spatio-temporal data and maps for the use of others.

4.4.7 Pseudo-object GIS

The least common geometry method discussed in the previous section is based on region objects and may be applied with every GIS which is able to handle such objects. One major disadvantage is, that the building of least common geometries and the related storage of information has to be performed in consecutive steps. This is necessary to re-combine the least common geometries to valid spatial units. Even minor changes concerning one spatial object require a complete re-building of the database. In practice it is almost impossible to use bi-temporal data (see chapter 3.3), including valid and transaction time.

In general, valid and transaction time may be simulated using the LCG method based on an attribute level versioning of a temporal database, as explained in chapter 3.4. However, the complexity of the spatial objects will increase dramatically.

The time-space composite shown in figure 4.27 may also be build with line segments instead of regions (amendment vector method), using a topological data structure (see chapter 2.3; e.g. arcs in Arc/Info). Additional line segments may be added any time after the completion of the database (see figure 4.29).

The workflow shown in figure 4.30 addresses the same steps as applied in the case of the least common geometry method. The details of the adaptations depends on the software package used. The approach is highly versatile, the line segments are geo-referenced and digitised in a GIS. Whether the line segments are also stored in the GIS as shown in the figure or exported and stored in a relational database management system depends on the GIS and the project goals. The most flexible treatment is to include the point and line segments in the relational database management system and store them together with the entity-relationship model of temporal changes.

The fields of application as well as the constraints for the implementation are similar to those related to the least common geometry approach. However, the more versatile and flexible data structure is an advantage of the design based on points and line segments. Whether this approach is more expensive than the least common geometry method depends on whether there is already a GIS in use, that uses topological structures. If yes, the additional effort can be minimized. If not, the cost of a GIS package which uses topological structures has to be balanced against the advantage of covering transaction time in the system.

5 Processing and analysis of spatio-temporal data inside a GIS

> "A crystal ball is perhaps the most important peripheral device to have available when implementing a TGIS."
>
> (Gail Langran 1993)[36]

Geographic Information Systems are distinguished from graphics and CAD software packages by the available analysis functions as well as the linkage of geographical and attribute data. The target of all analyses within a GIS can be summarized as the requirement of generating new information by theory-led combination of available data stocks (Strobl 1992: 48). The following chapter discusses basic analysis functions implemented into GIS. We may differentiate between spatial, thematic and temporal queries.

5.1 Common GIS analysis functions

In general, commercial GIS packages offer a broad range of different analysis functions, which can be used for the investigation of temporal processes. In addition, more complex software packages permit the programming of own analysis modules or the integration of macros for the automation of analysis steps.

The different methods of analysis can be grouped according to software-technical or content criteria. From a methodical or software-technical point of view the analysis categories query and measurement functions, statistical procedures, overlay and splitting techniques, geometric-topological operations as well as interpolation, network and modelling or simulation functions can be mentioned (see table 5.1).

Basically the majority of analysis functions may be applied to any geometric data structure. However, depending on the main focus of analysis either vector or raster based data structures may be preferred. Road networks for example should be represented in a vector environment. The underlying nodes and arcs constitute the basis for network analysis and optimisation. In raster systems such operations are far more complicated or sometimes almost impossible. If, by contrast, areal features such as diffusion are analysed, the raster data model suits much better (Saurer & Behr 1997:139).

---

[36] Cited after Luttermann (1994:9)

raster		vector
top-down		bottom-up
holistic view		elementary view
disaggregating	vs.	aggregating
image-oriented		object-oriented
pattern-oriented		structure-oriented
Interpretation		explicit modelling

Fig. 5.1 Perspectives of spatial analysis (Strobl 1992: 50)

Following the different approaches of conceptualising the real world presented in chapter 2.2.3 we may assign the two geometric data models to two complementary perspectives of spatial analysis (see figure 5.1). At the end of the day, the selection of a suitable data model must be based on project goals, subject and scale of analysis as well as data availability and data quality. Hybrid data models normally reserved to premium GIS packages increase the flexibility of spatial analysis procedures.

Regarding the content related analysis four model types for the description and analysis of spatio-temporal processes can be distinguished after Kilchenmann (1992:9; see figure 5.3). The *geographical model* asks for location parameters as well as the neighbourhood relations of the object or the region. While the *multivariate model* is interested in the features of a spatial object and the connections between developments of these features at a certain point in time, the *historical model* regards an object or a region at different points in time. By joining the three model types we receive a complex *multivariate historical-geographical model* of the spatio-temporal processes. It forms the basis for a multitude of applications concerned with the analysis of spatio-temporal data.

The model types mentioned can be assigned to four groups of analysis methods (Saurer and Behr 1997:135) which will be discussed in the subsequent sections.

1. *Location analysis:* Functions to determine, change, evaluate and analyse the spatial properties of an object.
2. *Attribute* analysis: Techniques to enable the handling and analysis of the attribute data linked with the spatial objects.
3. *Time series analysis:* Procedures to permit the documentation and investigation of modifications to the spatial or thematic characteristics of an object at different points in time.
4. *Process analysis:* Operations to enable the modelling of spatial and thematic modifications of an object in the past and present as well as a forecast of the future development.

Performing queries on a GIS database is essential to almost any GIS project. The queries may either be performed on data that are explicitly part of the GIS databases or on new data resulting from data analysis procedures within the GIS.

Table 5.1 Common GIS analysis functions

category of analysis	method of analysis
1. query and measurement functions	▪ query of location, length, area, perimeter, etc . ▪ query of assigned attribute values
2. statistical analysis	▪ uni- and bivariate methods ▪ multivariate methods ▪ spatial statistics
3. overlay/split	▪ Boolean Operators (cf. fig. 5.4): common combined occurrence (.AND.), common expansion (.OR.), one-sided exclusion (.NOT.) or mutual exclusion (.XOR.) ▪ Fuzzy-Logic: The procession of natural phenomena with Boolean logic is „too exact". Fuzzy logic suits better for the continuum character
4. geometric-topologic operations	▪ buffering ▪ neighbourhood ▪ interaction ▪ distance, diffusion
5. interpolation	▪ digital terrain models ▪ statistical surfaces ▪ point and areal interpolation
6. network functions	▪ best route ▪ best locations ▪ travelling salesman
7. temporal analysis	▪ temporal queries ▪ process analysis ▪ simulation

	raster	vector		
		point	line	polygon
spatial query	•	•	•	•
measurement	•	•	•	•
statistical analysis	•	•	•	•
overlay/split	•	(•)	(•)	•
buffering	(•)	•	•	•
neighbourhood analysis	•	(•)	(•)	•
distance operations	•	•	•	•
diffusion	•	(•)	(•)	(•)
Boolean operations	(•)	(•)	(•)	•
interpolation	•	•	–	–
network analysis	–	–	•	(•)
temporal query	•	•	•	•
process analysis	•	•	•	•
simulation	•	•	•	•

Fig. 5.2 Data structures and analysis functions

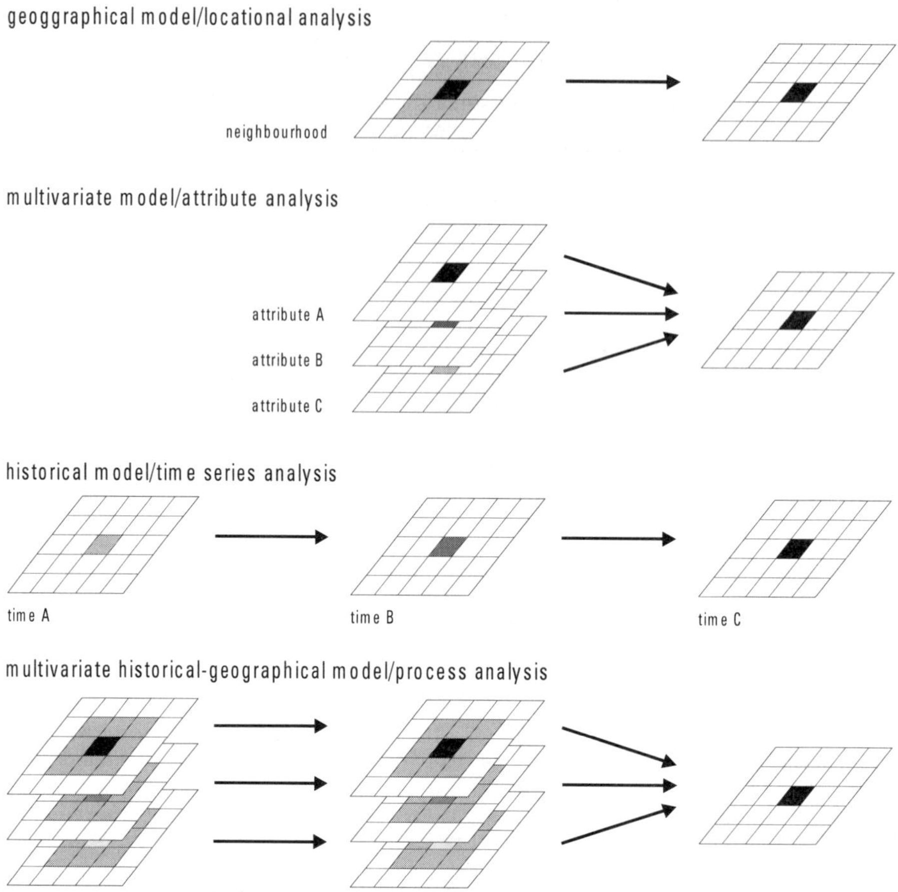

Fig. 5.3 Model types of spatio-temporal processes (based on Kilchenmann 1992: 9)

As mentioned several times in the preceding sections, the modelling of spatio-temporal processes is impeded by the restrictions of the layer oriented approach of common GIS packages. A clear improvement will in the long run only be achieved with the development and implementation of object-oriented GIS.[37] However, process analysis procedures or prognostic models will also in the future only be integrated in exceptional cases as standard components in commercial GIS packages.

[37] The advantages of object oriented spatial analysis are described by Ralston (1994).

5.2 Spatial and thematic queries

5.2.1 Spatial queries

Spatial queries are among the basic functions of any Geographic Information System. For this purpose an object is selected by clicking on the digital map with either mouse or puck. After input of a corresponding command, both, data that refers to the geometrical qualities of the chosen object (for example area, circumference, centroid coordinates, number of coordinate pairs that form the object) as well as attribute data assigned to this object via an ID variable are displayed. A GIS reduced to this functionality represents a simple inquiry system.

In addition, most vector GIS packages are able to aggregate objects temporarily. The resulting query then displays statistical parameters such as sum, mean, minimum and maximum or standard deviation for the selected subset of objects. The subset may be transformed into a new permanent object by uniting the selected objects. The attribute data sets associated with the selected objects are summarized in a single data set which is automatically linked with the new object.

5.2.2 Thematic queries

Unlike object-related queries, where information about a known object is filtered from the database, thematic queries are applied on the database level. Objective of such analyses is to identify spatial units that satisfy specific contents-related conditions. The queries are formulated in a database language (e.g. SQL) and are based on the connection of mathematical terms with Boolean operators (see figure 5.4). Depending on the functional scope of the respective GIS package very complex interrogations may be assembled.

Given the example of a cadastral map with objects (plots) described among other things by the variable “area“ the logical expression “area > 500“ leads to the selection of all lots whose sizes exceed 500 m². If the plots are characterized by a further variable “built_up“ with the values “true“ and “false“, the query “area > 500 AND built_up = false“ identifies all plots without buildings with a minimum size of 500 square meters. This synchronous examination of several different data layers is usually referred to as *multi criteria analysis*.

In raster GIS often reclassification techniques are used to perform thematic queries. Consider a land use map with different cell values for different land uses. If one wished to identify all built-up areas, an answer could be obtained by reclassifying the image into a new raster map. The new value of 1 could be assigned to all cells representing water in the original image and a value of 0 to those with a different land use.

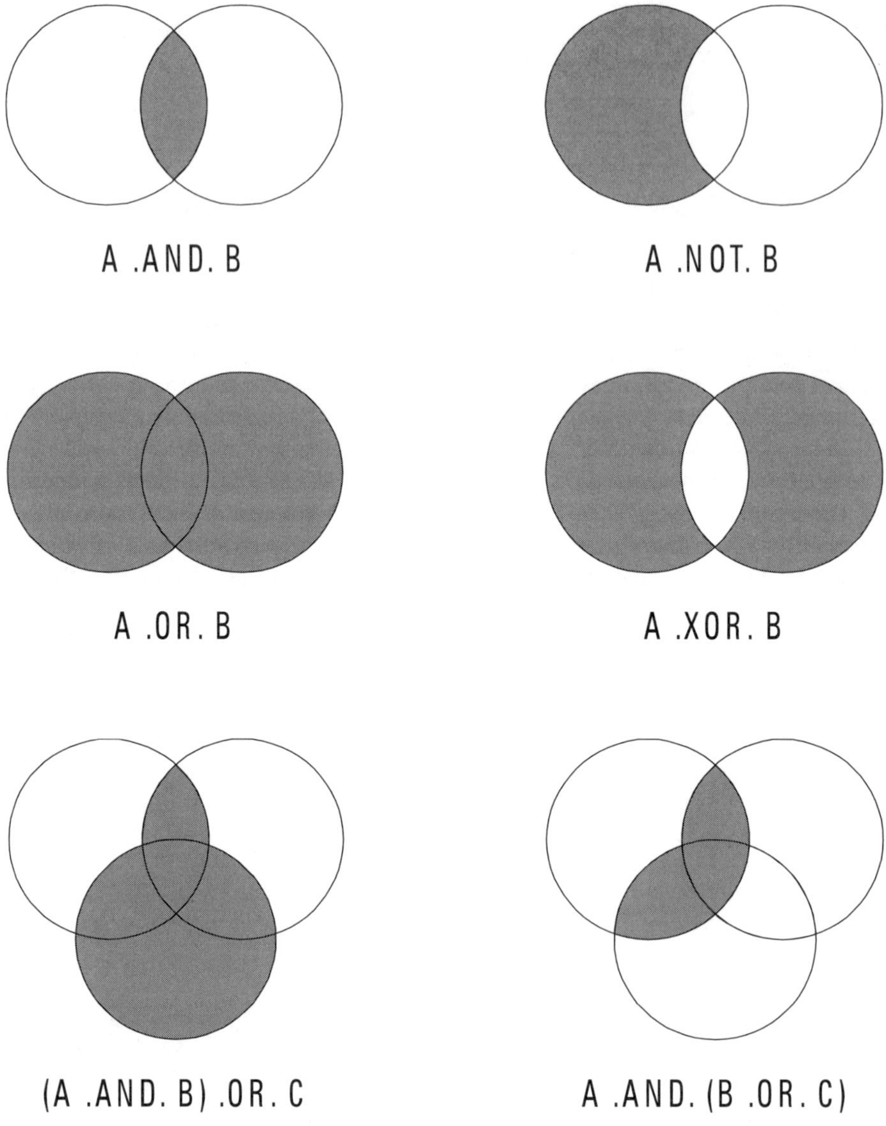

Fig. 5.4 Boolean operators

As the procession of natural phenomena with Boolean logic is often „too exact" fuzzy logic techniques are used instead. These procedures are suited better for the continuum character of natural entities.

5.2.3
Statistical analysis

As a rule, GIS packages offer standard data base operations to perform arithmetic calculations or classifications. For example the variable "area" automatically provided by the GIS may be used together with the „number of inhabitants" taken from a related attribute database to calculate the "population density" of administrative units. The functions may be grouped into functions that do not include geometric data, and those that explicitly use geometric or topologic data structures for the computations (geo-statistics). The former procedures are used to calculate global or categorical statistical parameters (e.g. average values, standard deviation) or to classify and condense the data. In contrast, the latter techniques are used to investigate the relations between the features of objects with consideration of the spatial reference (e.g. kriging).

In terms of dimensionality we must differentiate whether only a single pattern is under investigation (univariate), or whether more than one pattern is involved (multivariate). "The essential difference is that in the univariate case the only possible 'explanation' of observed pattern is in term of some overall spatial 'trend' across the locations, or possibly through a tendency for neighbours to 'move together', i.e. to be 'spatially autocorrelated'; whereas in the multivariate case additional 'explanation' is available via other 'spatially correlated' attributes measured at each of the locations" (Bailey 1995:18-19).[38]

In many standard GIS packages there is a lack of functions and methods for analysing structural or spatial patterns as well as spatial correlations among the patterns of distribution. In order to perform advanced analyses it is often necessary to combine the GIS with specialized statistics software packages.[39]

5.2.4
Overlay and split

Overlaying and splitting techniques are fundamental analysis functions in any GIS. Based on the principles of a layer GIS spatial information is combined by using spatial attributes (location). Overlay and splitting procedures emulate the functioning of a cookie cutter. The objects in a layer are stamped with the objects of another layer. During the division of objects attribute data can be passed on. This passing on can be adjusted by parameters, e.g. with respect to the size of the areas. Thus, in principle a disaggregation of attribute values to split areas is possible. Other analytical functions (e.g. weighting) can be integrated into the splitting process. However, this process gives only valid results under the assumption of

---

38 Cf. Bailey (1995) for a structured overview of potentially useful statistical spatial analysis techniques in relation to Geographic Information Systems.

39 The AI-GEOSTATS Website [http://www.ai-geostats.org/] is an excellent resource for finding information on available freeware and commercial software packages commonly used in spatial data analysis and spatial statistics.

equal distribution. In a vector oriented GIS three basic types of overlay may be distinguished:

- overlay of point and polygon:
  The overlay of point and polygon is trivial in geometrical terms. The resulting points are attributed with the information associated to the containing polygons.
- overlay of lines / overlay of line and polygon:
  In this case two basic forms may be distinguished: First, a polygon can be overlaid and split by a line. The information of the original polygon is assigned to the resulting polygons in respect to certain rules set by the user. Usually the attribute data is split proportionate to the area of the original polygon. Second, a line may be split by another line or a polygon. The resulting line segments are either associated with attributes of the splitting line or polygon, or the line segments may be assigned with proportional values of the original line.
- overlay of polygons:
  Splitting two polygons may either result in the proportionate split of attribute data of the original polygon, or in the assigning of attribute values of the cookie cutter polygon to the resulting polygons.

5.2.5 Geometric-topologic operations

The analysis of topologic neighbourhood relationships ranks among the most frequently carried out analyses. It can be applied equally on raster and vector data. Common neighbourhood operations include the calculation of discrete distance corridors around points, lines or polygons, called buffering. Beyond mere visualization, the produced buffer zones can be included in further calculation and analysis steps. Analyses in the field of distance and diffusion calculations (e.g. forest fires, disease control) are executed predominantly with raster systems; they are, however, also feasible with vector data. As demonstrated in figure 5.5 buffers are used to establish critical areas for analysis or to indicate proximity or accessibility conditions. If the concept of gradation is required, a sequence of buffers at selected increasing distances must be created (Laurini & Thompson 1996:279-280).

In general, neighbourhood operations put a spatial object in relation to their neighbouring objects and enable the generation of new information from the environment information. Local neighbourhood operations are mostly applied in raster based GIS: the relation of pixels to their immediate neighbours is analysed. In the narrow sense only pixels with adjoining edges count as neighbouring pixels. A more extended comprehension of neighbourhood also counts those pixels who adjoin only with their corners. Thus, according to the favoured point of view a pixel may have four or eight neighbours. A typical example is the derivation of a slope map from absolute elevation values. Each cell or pixel of the elevation map is characterised by one altitude value. In order to estimate the slope value the GIS calculates the elevation variations of the four or eight neighbouring cells. As each cell of the resulting slope map can only hold one slope value the user can decide whether to take the minimum, maximum or a median slope value.

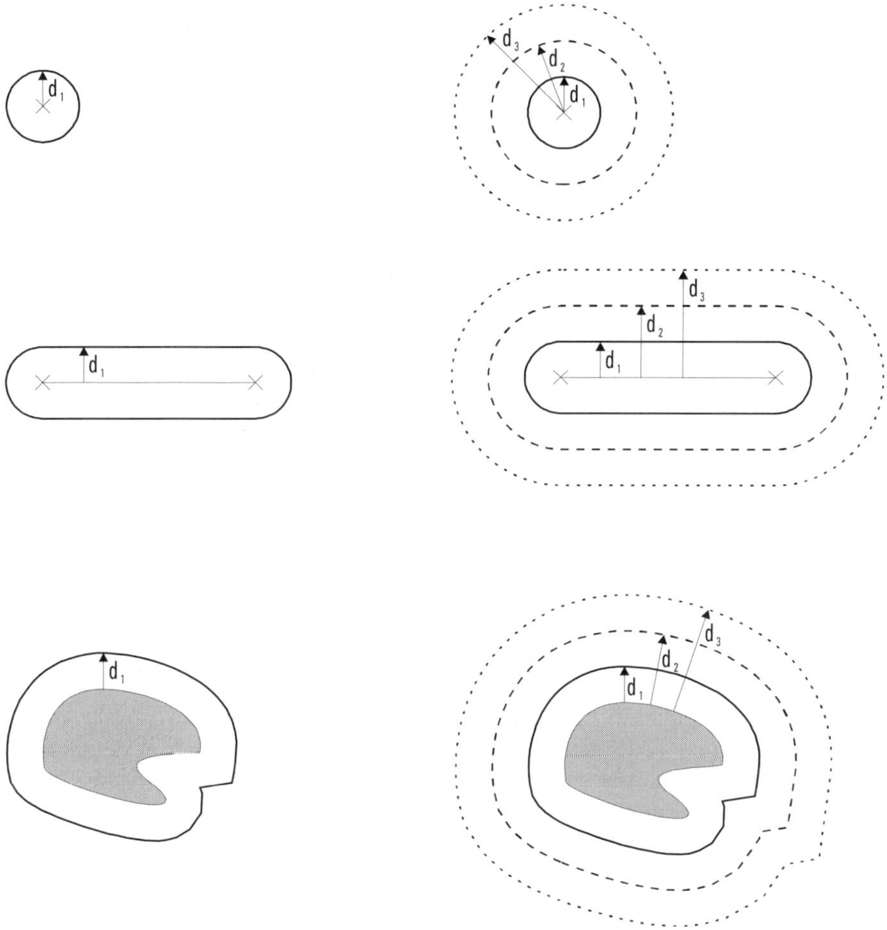

Fig. 5.5 Buffering in vector GIS

5.2.6 Interpolation

The problem of spatial interpolation can be simply described as follows: Given a set of spatial data (either as discrete points or for sub-areas), the function that will best represent the whole surface and that will predict values at other points or for other sub-areas must be determined. The main applications of different spatial interpolation methods are in isoline mapping or in the transformation of polygons into point data (e.g. Thiessen-polygons).

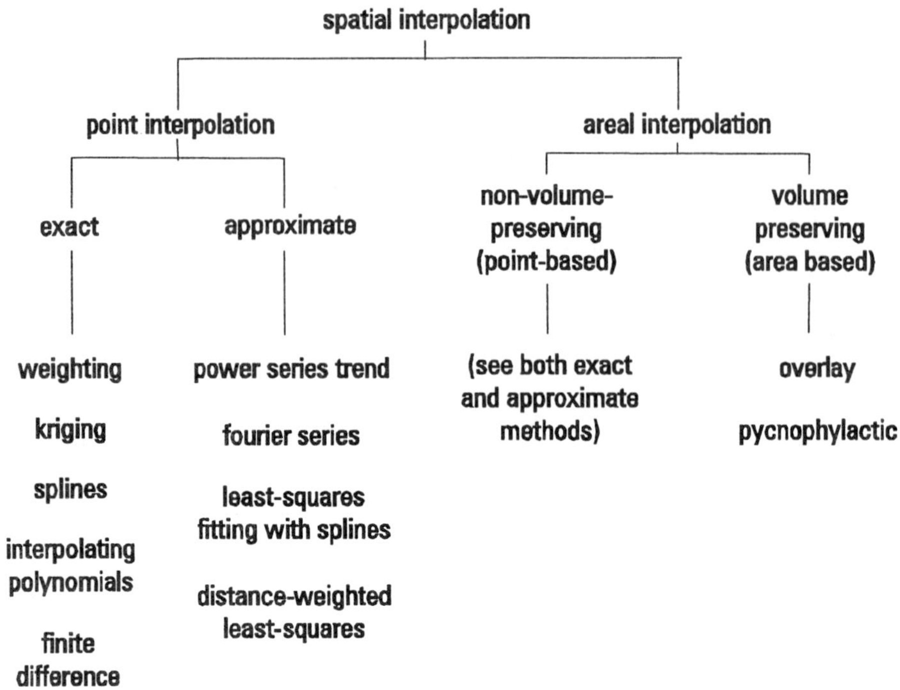

Fig. 5.6 Interpolation methods (adopted from Lam 1983:130)

The various interpolation methods may be differentiated into two basic groups. *Point interpolation* deals with data collectable at a point, such as temperature readings or elevation, whereas *areal interpolation* deals with data that are aggregated over a whole area (e.g. population counts per census tracts). Maps of the first type are usually referred to as isometric, those of the latter type as isopleth (Lam 1983:129). Point interpolation methods may be further subdivided into exact and approximate methods according to whether they preserve the original sample point values, areal interpolation methods are differentiated according to whether they preserve volume or not (see figure 5.6).

Interpolation methods are often used to create digital terrain models or statistical surfaces. While the former is used for the visualization of terrain surface and the derivation of relief-referred measures (e.g. slope inclination, exposure), the latter is a method to visualize object attributes with exact spatial reference in 3D (see chapter 8).

5.2.7
Network functions

Procedures to analyse network structures are not always part of the analytical toolbox of standard GIS packages.[40] The vector based algorithms are used to solve a number of classic optimisation problems such as nearest neighbour, shortest path or travelling salesman. They usually focus the search for the shortest, fastest, lowest-priced or otherwise favourable connections in a network. As a mathematically correct solution ("travelling salesman") cannot be calculated with sufficient performance for large networks, usually approximation solutions are used. Often, the optimisation result may be expressed using temporal data such as saved driving time.

Apart from the pure path search, functions exist, which analyse the network topology with integration of attribute data. In systems, which depict networks topologically, it is possible to implement complex network pursuit functions. Networks may also be used to represent migration processes and other spatial interactions.

5.3
Temporal queries

Basically, temporal queries are based on the same principles as spatial queries. Thus, the Boolean operators described in section 5.2.2 may be adopted to the temporal domain. This similarity is illustrated in table 5.2 referring to some possible spatio-temporal operators for objects and events. Applied to linear time intervals (see figure 5.7) this correspondence becomes even more obvious.

Table 5.2 Spatio-temporal operators (Wachowicz & Healey 1994:109)

spatial operators	temporal operators
disjoint	early/late
meets	meets
equals	equals
contains	during
covers	starts/finishes
overlaps	overlaps
intersects	intersects

[40] Usually toolboxes such as the ArcView Network Analyst are available as add-ons.

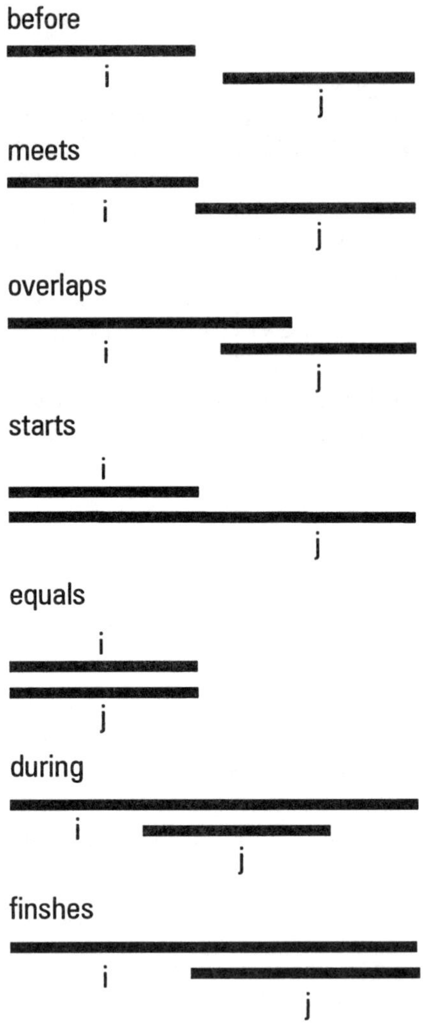

Fig. 5.7 Relations between linear time intervals (adopted from Frank 1998:41)

Computer scientists have laid the foundations for temporal queries by developing a large number of languages such as TSQL, a superset of the structured query language used to retrieve data from relational databases. The functions provided by TSQL include comparing time intervals, computing the intersections between time periods, analysing time discontinuities, aggregating over time periods and choosing the length of a time interval. Since TSQL is an extension of SQL, it provides an obvious starting point to integrate time-based queries into the majority of existing GIS packages. The objects in the GIS need to be time-stamped with a begin- and end-time; single points in time are represented by a zero duration time interval, i.e. begin-time equals end-time (see table 5.3).

Table 5.3 Time-stamp- and event-based TSQL operators (derived from Claramunt & Thériault 1995:35)

(a) time-stamp-based operators

BEFORE	[a,b] BEFORE [c,d]	$b<c$
AFTER	[a,b] AFTER [c,d]	$a>d$
DURING	[a,b] DURING [c,d]	$(a\geq c)$ AND $(b\leq d)$
EQUIVALENT	[a,b] EQUIVALENT [c,d]	$(a=c)$ AND $(b=d)$
ADJACENT	[a,b] ADJACENT [c,d]	$(c-b=1) \mid (a-d=1)$
OVERLAP	[a,b] OVERLAP [c,d]	$(a\leq d)$ AND $(c\leq d)$
FOLLOWS	[a,b] FOLLOWS [c,d]	$(a-d=1)$
PRECEDES	[a,b] PRECEDES [c,d]	$(c-b=1)$
STARTS	[a,b] STARTS [c,d]	$a=c$
FINISHES	[a,b] FINISHES [c,d]	$b=d$

[a,b] and [c,d] symbolize the two periods begin- and end-time.

(b) events-based operators

EVOLUTION	[InstanceA1] CHANGES TO [InstanceA2]	$A1=A2$
SUCCESSION	[InstanceA] REPLACED BY [InstanceB]	$A\neq B$
PRODUCTION	[InstanceA] PRODUCES [InstanceB]	$A\neq B$
REPRODUCTION	[InstanceA] GENERATES [InstanceB]	$A\neq B$
TRANSMISSION	[InstanceA] TRANSMITS [InstanceB]	$A\neq B$

The symbol $A1=A2$ means that the operator is restricted to two versions of the same entity, while $A\neq B$ implies two different entities.

The time-stamp-based operators (a) are used to compare time periods. They do not provide an explicit way to model and distinguish evolution, succession, production, reproduction or transmission processes. Indeed, time-stamp-based operators use time measurements to compute relative positions between potentially independent entities as opposite to the topological operators that link interdependent entities (b). In this case "instance" has the meaning of a generalized data type in the temporal domain (Claramunt & Thériault 1995:36; see figure 5.8):

- The CHANGES TO operator returns "true" when InstanceA2 represents a mutation immediately following InstanceA1.
- The REPLACED BY operator returns "true" when InstanceB is an immediate successor of InstanceA. Both instances may survive the replacement with Instance A being permuted to another position. The succession process is encoded as a *1:n* or a *n:1* relationship. If a relocation process of the *n:m* type is required, this can be modelled using a *n:1* followed by a *1:n* relationship applied to an intermediate virtual object.

- The PRODUCES operator returns "true" when InstanceA is the producer of InstanceB. The production is modelled as *1:n* where the producer and the product are entities from different classes.
- The GENERATES operator returns "true" when InstanceA is an immediate ancestor of InstanceB. The reproduction is also modelled as 1:n. However, producer and product must be different entities of the same class.
- The TRANSMITS operator returns "true" when InstanceB has its attributes modified by some kind of contact with InstanceA. The main difference with the production and reproduction processes is that InstanceB already existed prior to the transmission process.

The similarities between spatial and temporal analysis functions also apply to other techniques such as temporal buffering or the computation of temporal isolines (e.g. isochrones). Other operations explicitly include temporal information, such as routing algorithms or diffusion calculations.

5.4 Process analysis and simulation

The analysis of spatial or geographical systems and processes depends on the formulation of dynamic concatenations of otherwise isolated operators and entities. There is a classic sequence of system analysis, modelling and simulation which aims at the explanatory and prognostic analysis of spatio-temporal processes. The process of designing a model of a real world system and conducting experiments with the model for the purpose of describing, explaining and predicting the operation of the system is called simulation. It is fairly obvious that both time and space are inseparable components of any process analysis or simulation.

Aside from mock-up hardware models several basic types of simulation models are recognized:

- *Conceptual models* are usually flow diagrams drawn on paper or the computer screen which can be used to organize ideas.
- In *mathematical models* the key features of a system are translated into the symbolic logic of mathematical formulas. In general, computer simulation models are mathematical models.

The development of computer simulation models in geography and other social sciences is a consequence of the so-called quantitative revolution, which generally encouraged the use of systems theory and modelling techniques. The interest in the dynamic elements of physical and human environments stimulated the use of simulation models and other comparable techniques which offer more than static snapshots of spatial systems. In this context models may be classified as either deterministic or probabilistic. *Deterministic models* are based on the concept that the subject to be analysed can be described exactly using mathematical relationships. Therefore deterministic models are almost exclusively restricted to physical applications. On the opposite, *probabilistic models* are based on the notion that the subject under study may be subject to random influences. They equations include information on expected probabilities of certain events and processes occurring. This model type is often used to simulate human decisions. In a computer simulation model elements of both deterministic and probabilistic modelling strategies can easily be combined.

The model building process follows the concepts of abstraction and conceptualising described in chapter 2.2. In a computer simulation model the relevant features of a particular pattern or process are represented by a set of variables. One or several procedures are introduced to calculate a new set of values for the variables, taking into consideration the dynamics of the system. These procedures must then be translated into a stand-alone computer program or a sequence of commands in a macro language.[41] During the operation of a computer simulation, the calculations indicated by the equations of the model are performed repeatedly and/or recursively to represent changes over space and time (Maguire 1989:111-112).

Simulation allows the temporal dimension of processes to be compressed and extended, so that long-term processes (e.g. continental drift or population development) and short-term processes (e.g. landslides or wildfires) can be investigated conveniently. Another major advantage computer simulations have to offer is, that key parameters of the model may be altered deliberately. By re-running the simulation competing theories can be tested and evaluated. Once a valid set of parameters is found, the model may be used to simulate future events or to predict potential developments (Maguire 1989:114).

Chapter 7.5 demonstrates the basic procedures of process analysis and simulation by presenting a GIS-based wildfire prediction application.

[41] e.g. Arc/Info AML, ArvView Avenue or Mapinfo's Mapbasic

6 The visualisation of spatio-temporal processes

> "Virtual reality is no reality at all. The weight on mankind of time and space, of physical surrounding and history – in short, of geography – is bigger than any earthbound technology is ever likely to lift."
>
> (The Globe and Mail 1974)[42]

An important aspect of Geographic Information Systems deals with the functionalities used to visualize spatial data. Unsurprisingly this task is usually accomplished by applying methods and techniques provided by cartography, as maps are both, the means and the product of geographical analysis. Recently, many new impulses from the emerging field of visualisation in scientific computing (ViSC) continue to have a strong influence on the methods used to visualize spatio-temporal data. This chapter deals with some basic insight on the visualisation process as well as with classic and modern approaches of displaying temporal data on paper or a computer screen.

6.1 Basic concepts of spatio-temporal visualisation

From a computer cartography point of view, visualisation is considered as a process for putting images into human minds. For example, the shape of a mountain is poorly conveyed by contours, but it can be better presented by rotating a 3-dimensional block diagram. Considering air-flows over a patch of terrain, a 3-dimensional display as well as animation to show true pattern of directions or speeds of flow are needed. In social-economic studies, presenting movements of people in an area may need the ability to generalise individual movements. The pattern of growth of an urban area may be difficult to analyse using static paper maps, and it may be better understood if one uses computer animation to show changes through time effectively. Following Kraak (1999a) three basic objectives and purposes of visualisation may be distinguished:

- First, visualisation may be used to *present* spatial information. The results of spatial analysis procedures may be displayed in printed maps or multimedia applications in order to answer questions such as "where is", "what is?", or "what

[42] August 1, 1994, p A1, cited after Taylor (1999:325)

belongs together?". The science of cartography offers an abundance of rules and examples how to create well-designed and understandable maps for a wide audience.

- Second, visualisation techniques may be used to *analyse* a set of data. This includes basic GIS procedures such as overlay, where two separate layers of information are combined in order to understand the interrelations. Required are functions to access map components and functions to process, manipulate and summarise that information. Questions like "what is the best site?" or "what is the shortest route?" may be answered.
- Third, visualisation may be used to *explore* huge amounts of raw data. These techniques are generally referred to as data mining; they are often applied prior to actual analysis procedures in order to identify datasets that are appropriate for the problem under investigation.[43]

As shown in figure 6.1 the three strategies may be arranged in a cube, where the axes represent the dimensions of audience (public–private), interaction (low–high) and nature of the data (known–unknown).

Visualisation changes the user's role from passive to active. "Maps are changing from being final products presenting spatial information to interim products that facilitate visual thinking" (Kraak 1998, cited after White 1997: 6). Depending on the position along the diagonal inside the "map use cube", a key feature of visualisation is the ability of the user to change the information on the screen interactively, for example the mapped variable, the temporal aspect of the map, or the cartographic representation of objects on the map. Most of the benefits provided by visualisation techniques are related to dynamic elements, such as zooming or panning, moving the cursor in order to reveal hidden information (e.g. images or texts), or clicking a button to change variables or colours.

One might assume that a Geographic Information System, specifically designed for the storage, manipulation and display of geographic data, would provide a useful environment to explore the data visually. Unfortunately in current GIS, the number of graphical primitives under the direct control of the user is small. On most systems, only 2D or 2.5D at best is supported, along with a choice of colour or line styles and fill patterns. Thus, the following ways of visualising time in a GIS may be listed:

- mono-temporal layer for one time step
- multi-temporal layer overlay for different time steps
- time series mapped with graph
- time series mapped with polar diagram
- 2,5D surface with temporal information mapped as z
- 2,5D surface with elevation, temporal data mapped on the surface
- space-time path
- space time composite as 2D map
- space time composite as 2,5D block diagram (see figure 3.6)

[43] Cf. e.g. Tucke 1994 on exploratory data analysis.

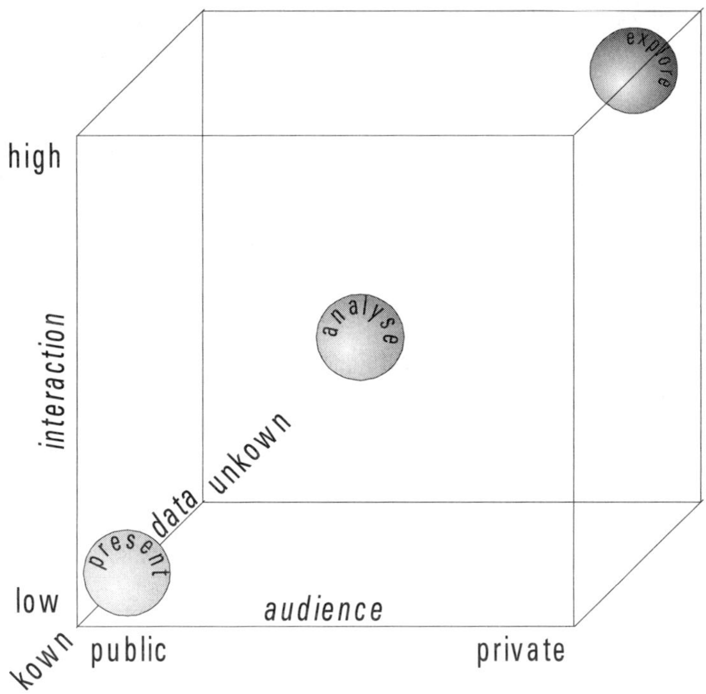

Fig. 6.1 The map use cube (adopted from MacEachren & Taylor 1994)

6.2 Displaying time on a map

Maps are composed of different kinds of symbols which specify points, lines and areas. Visual differences among map symbols convey information to be extracted by the map-reader. In GIS visualisation, these symbols are defined by graphical variables such as location, colour, size, shape, spacing and orientation. Papermaps are the most commonly used tools to present spatial data and distributions. However, their use and production are significantly restricted by the problems listed below:

- Maps are limited to two dimensions. As the real world is never two-dimensional, maps must show data in the third (or fourth) dimension by projecting them onto a flat surface. Obviously, this operation will give a distorted impression of spatial distributions, in particular those at the scale of the globe.
- Maps are static and can show interactions, flows or changes through time only by using oblique methods such as symbolic representations.
- Maps have difficulty to present uncertainty in the data. Very often this gives a false impression of accuracy.

The data displayed in thematic maps is always tied to a certain point in or period of time. This temporal aspect plays an important role in the analysis of the maps by the user, especially if different time slice maps of the same region are compared. The temporal components become even more crucial when attribute data are available for different points in time and must be represented in a single map. However, understanding the temporal phenomena from a single map will depend on the cartographic skills of both, the map maker and the map reader, since these maps are usually very complex. An alternative is the use of single maps each representing a different time slice. Comparing these maps will give the user an idea of change (Kraak 1999a).

Thus, there are four basic ways to display temporal changes on a 2D map, by using either choropleth maps, temporal diagrams, change indicators or a combination of the three. The technique of representation again depends on four factors (Olbrich et al. 61-63):

- Properties of the data: continuous data may be displayed using line graphs, while discrete data can only be represented by bar charts.
- Number of spatial units: the larger the number of spatial units, the more simple the map must be constructed. Diagrams must not fall short of a certain size. Maps with a multitude of spatial units may be constructed using categories such as increase, stagnation or decrease.
- Number of variables and number of time slices: The constraints for large numbers of variables and/or time slices apply by analogy with those mentioned for a large number of spatial units.
- Research hypothesis: Depending on the questions to be answered the map may either stress the spatial patterns at the different points in time, or the changes of individual spatial units over time.

In general, there are four basic temporal constructs that may be included on a map (Muehrcke 1978; Halls & Miller 1996):

- *Qualitative change:* what has changed?
- *Quantitative change:* by how much did it change?
- *Composite change:* the process of change itself.
- *Space-time ratio:* descriptors used to express the relations between space and time.

Both, the requirements for cartographic representation listed above and the visualization capabilities of present soft- and hardware, result in fundamental display mechanisms (Grelot & Chambon 1984; Halls & Miller 1996):

- *Change data:* additions to a base map in form of diagrams or texts.
- *Static symbols:* a thematic map with temporal symbols depicting dates, rates, routes or developments.
- *Time sequences:* a series of time slice maps.
- *Animations:* space and time are scaled to model the metamorphosis of the area and/or attribute data under study (see section 6.4)

Figure 6.2 compiles a summarizing view on the cartographic representation of spatial objects. In a technical sense, every entity of the real world may be represented on a map by using points, lines, polygons or solids.

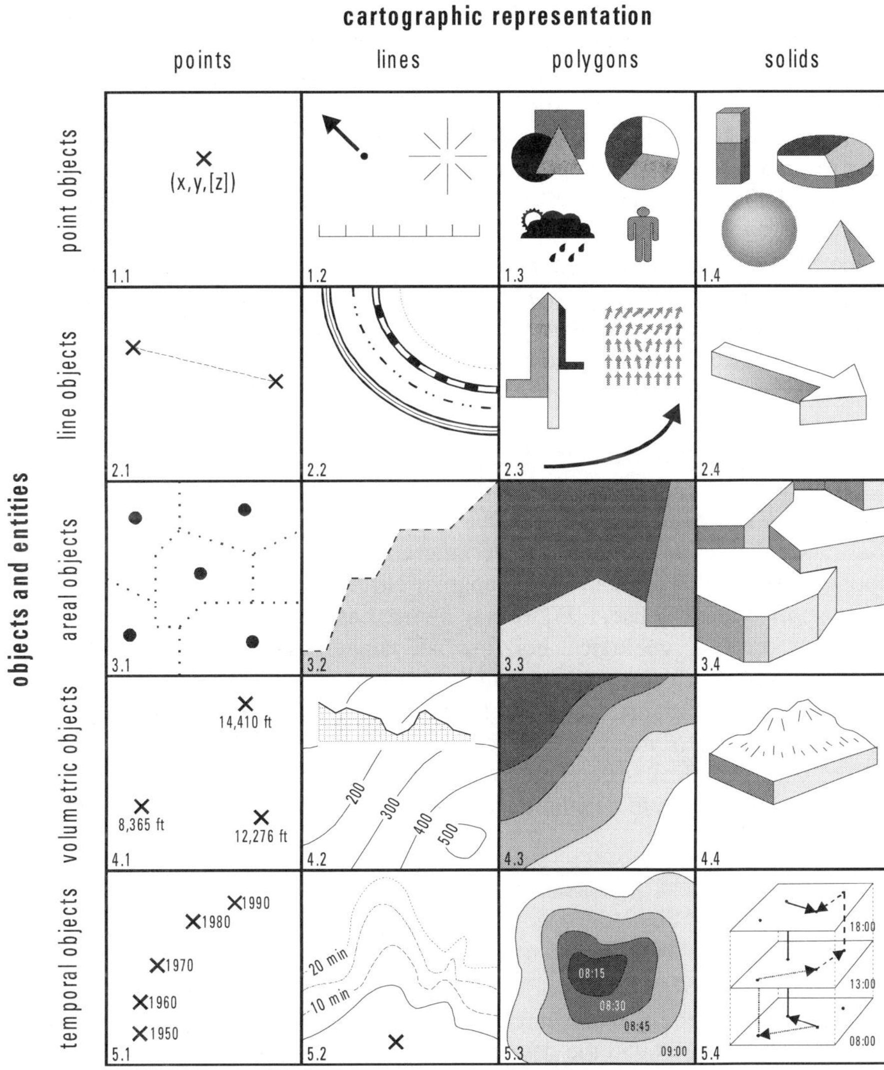

Fig. 6.2 Objects and their cartographic representation (adopted from Ormeling & Van der Schanz 1997:96)

6.3 Multidimensional GIS

To pick up the thread from chapter 2.5 the kind of data which can be processed and visualised in a GIS may be classified in five dimensions. Thus we may distinguish 2-, 3-, 4- and 5-dimensional Geographic Information Systems. For the purpose of presentation only two dimensional data can be mapped directly. A third dimension may be added with an additional z-value for each x,y coordinate, e.g. in order to model elevation. Perspective views of 3D data sets are called 2,5D presentations. The fourth dimension is time; the fifth dimension is the attribute data value.

The term digital elevation model or DEM is frequently used to refer to any digital representation of a topographic surface (i.e. a 3D GIS). It is, however, most often used to refer specifically to a raster or regular grid of spot heights. If the model is enhanced to contain additional information such as ridges, cliffs, valley lines and rivers, then the model becomes known as a digital terrain model (DTM). A DTM may be understood as a simple digital representation of a portion of the earth's surface. As overhanging cliffs and faults are relatively rare in nature, topographic surfaces are most often represented as "fields" (i.e. simply connected surface models, having unique z-values over x and y). In this sense a DTM is a 2.5D rather than a 3D model. However, there are now software packages that can cope with data in this form and can recognise cliffs and faults.

In a more general sense, a DTM may be used as a digital model of any single-valued surface (e.g. geological horizons, air temperatures or population density, see figure 6.3). Such three-dimensional representations of attribute data are usually referred to as statistical surfaces. By draping two-dimensional features or image data on three-dimensional surfaces an attribute dimension may be added to a DTM or a statistical surface model. If several time slices of attribute data are available it is possible to calculate intermediate frames needed for the creation of an animation.

6.4 GIS and Multimedia

The rapid and on-going achievements made in the different areas of information technology have facilitated the development of multimedia systems and hence the use of multimedia in spatial analysis and modelling. Multimedia technology encompasses textual data, graphics, hyperlinks, images, sound, animations and video. Examples of this technology include various encyclopaedias or atlases on CD-ROM. The application of multimedia makes it possible to communicate with a computer through almost all senses. As a result, computers become more user friendly and easier to use.

Spatial multimedia technology has been applied in two ways, firstly in existing GIS packages supporting multimedia and secondly, in existing multimedia hard-

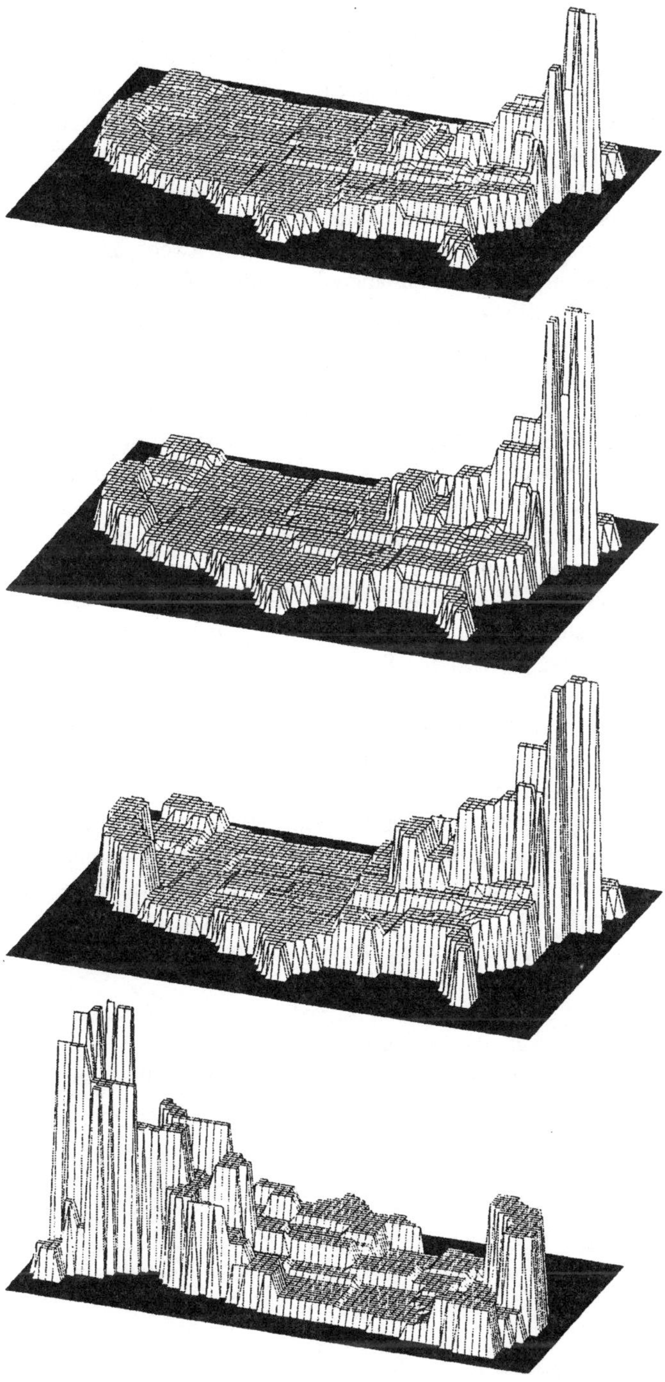

Fig. 6.3 Statistical surface: Population density in the USA, 1900, 1930, 1960, 1990

and software designed to handle spatial data. However, the full integration of new data types into an existing GIS has not yet been realised. Recent studies have aimed at using the potential of multimedia to integrate data of different kind, time and spatial scales. The use of multimedia technology for spatial analysis is basically considered as a tool for visualisation, simulation, communication and exploration of complex spatial data. The underlying notion is that making use of digital video and sound in one way or the other will contribute to better understanding of spatial phenomena and facilitate communication between and amongst groups of experts and users and therefore enhance the quality of decision making.

Multimedia animations for representing temporal data can be derived from static information or time layers. If only a few images need to be viewed in sequence, so-called animated GIFs provide a simple solution. Using a standard graphics software package or a shareware tool downloaded from the Internet, a series of images files in the GIF format can be joined in order to create an animation. The user needs to select the image files to be included, the order in which they are viewed, and how long each frame is displayed for.

User interaction can be increased by using movie file formats such as MPEG, AVI, RealVideo or Quicktime. These formats offer the user a set of controls similar to a VCR, such as play, pause or stop. This enables the user to view the entire animation, or to step into single frames in order to look for objects or patterns of special interest. The QuicktimeVR movie format even allows the user to move around and zoom into the scene.[44]

The Internet provides a revolutionary way of sharing geographic information The opportunities that unfold when integrating GIS with multimedia and WWW have caught the attention of GIS suppliers. Their reply has been the development of interactive mapping tools accessible on the Internet.[45] Moreover, the Internet enhances data visualization techniques in the sense of a dynamic or real-time map: Thematic data such as readings from a net of weather stations may be updated and displayed regularly and rapidly, and accessed from anywhere. Adapting to the concept of a client-server network many GIS suppliers have employed three key components:

- software that extends the functionality of a web server to accommodate HTML-compliant, interactive vector-based mapping,
- browser plug-ins that handle graphics downloaded by the Web/GIS server, and
- web mapping development tools.

---

44 A sample animation is included on the CD-ROM.

45 E.g. MapGuide by Autodesk, GeoMedia WebMap by Integraph, MapObejcts by ESRI or Pro-Server by MapInfo.

7 Examples

„Nothing in the world is purely spatial or purely temporal; everything is process“
J.M. Blaut (1961:2)

The examples given in the following sections are intended to be representative for the various fields of application of time-integrative Geographic Information System. Nevertheless readers should seek out applications that interest themselves, that deal with their own area, or that connect to their disciplinary background.

7.1 Building a historical GIS

For many countries historical census data is available for more than two hundred years. In addition there are large amounts of historical data collected during the last decades of quantitative historical research. This valuable pool of information kept in data archives around the world is rarely used according to its potential (see figure 7.1), as two constraints are imposed on such data:

- In many cases it is almost impossible to derive time series analyses from such databases because of breaks in the continuity of the territorial units used to assign the corresponding data. Often time series of statistical data are used, assuming by mistake that no change took place in the objects of research. Especially during industrialization most European countries underwent a process of administrative reorganisation with considerable territorial changes. In addition, the majority of central European countries look back on a eventful history of far-reaching territorial reforms related to nation building processes and wars.
- Historical data is rarely analysed using the inherent spatial dimension of the data. This is due to the lack of spatial analysis methods for historical data with their methodological impediments. Furthermore the spatial perspective is not widely included in the tools provided by traditional statistical analysis packages.

As a consequence, data for different time slices have to be aggregated or disaggregated to homogenise the impact of changing administrative units. If this is not possible, at least the mapping facilities of the GIS may be used to explore underlying patterns of a dataset which otherwise may be hidden beneath the hay stack.

The quantitative database of the Belgian municipalities, which covers the time span between 1796 and 1970 is an example for such an approach: "For all 3,000 municipalities that ever existed in this country between 1796 and 1970, the essential information on amalgamation/de-amalgamation and on boundary corrections was collected" (Vanhoute 1994:163). With this information a set of administrative maps representing every possible state and change was created, based on a modern municipality map of 1970. In the case of Belgium the municipalities can be used as a set of "natural" least common geometries because they have not changed considerably during the last 200 years. All administrative changes can be reconstructed from the municipality level. The system is based on a "concordance table" which stores the information on how to recombine separately stored parts of the maps, that represent stable states between two changes.

The approach is somewhat similar with the least common geometry approach presented in chapter 4.4 gaining from the advantage of relatively stable municipal boundaries as least common geometries. For other countries with a more complex record of changes of both, the administrative division and the administrative borders, a more versatile approach must be adopted.

	1	2	3	4	5	6	7	8	9	10	11
1796	•										
1801	•		•	•							
1806	•		•	•							
1816	•		•	•	(•)						•
1818	•		•	•							
1820	•										
1830	•	•	•	•	(•)						•
1846	•	•	•	•		•		•	•	•	•
1856	•	•	•	•		•		•	•	•	
1866	•	•	•	•	•	•		•	•	•	
1880	•	•	•	•	•	•		•	•	•	
1890	•	•	•	•	•	•		•	•	•	
1900	•	•	•	•	•	•	•	•	•	•	
1910	•	•	•	•	•	•	•	•	•	•	
1920	•	•	•	•		•	•	•	•	•	
1930	•	•	•					•	•		
1947	•	•	•		•			•	•		
1961	•	•	•	•	•		•	•		•	
1970	•	•	•	•	•	•	•	•		•	

1 Total population, 2 households, 3 sex, 4 civil status, 5 age groups, 6 origin, 7 foreign nationalities, 8 housing (1930 towns with population <10,000), 9 language used (1920 towns with a population >2,000), 10 education, 12 religion

Fig. 7.1 Content of Belgian densus data (Vanhaute 1994:169)

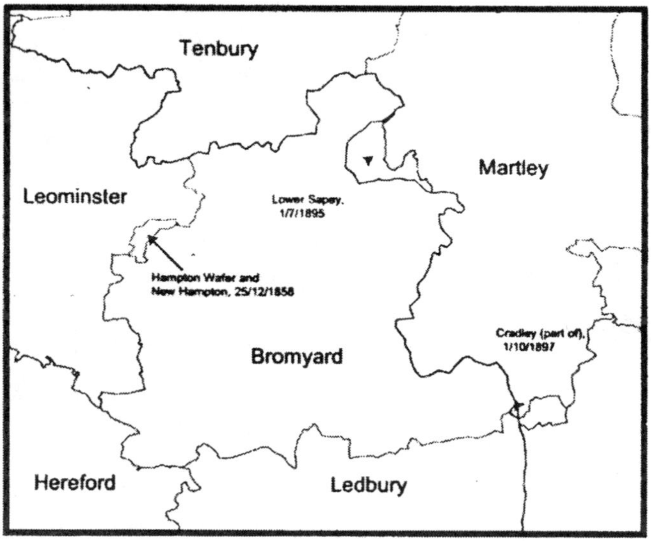

Bold lines represent the county boundary

0 5 10
Scale (kms.)

Areas transferred from Bromyard.

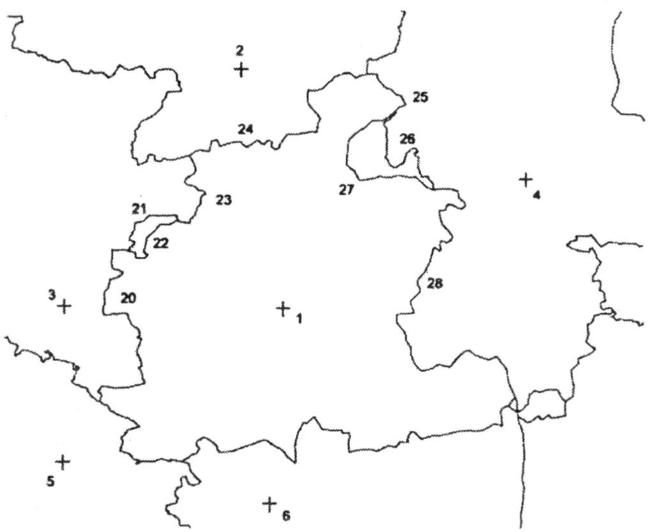

Crosses represent label points
Numbers are ID numbers on the point and arc attribute tables shown in 2c and 2d

Fig. 7.2 Digitization of historical boundaries for the Great Britain Historical GIS (Gregory & Gilham 1998:11

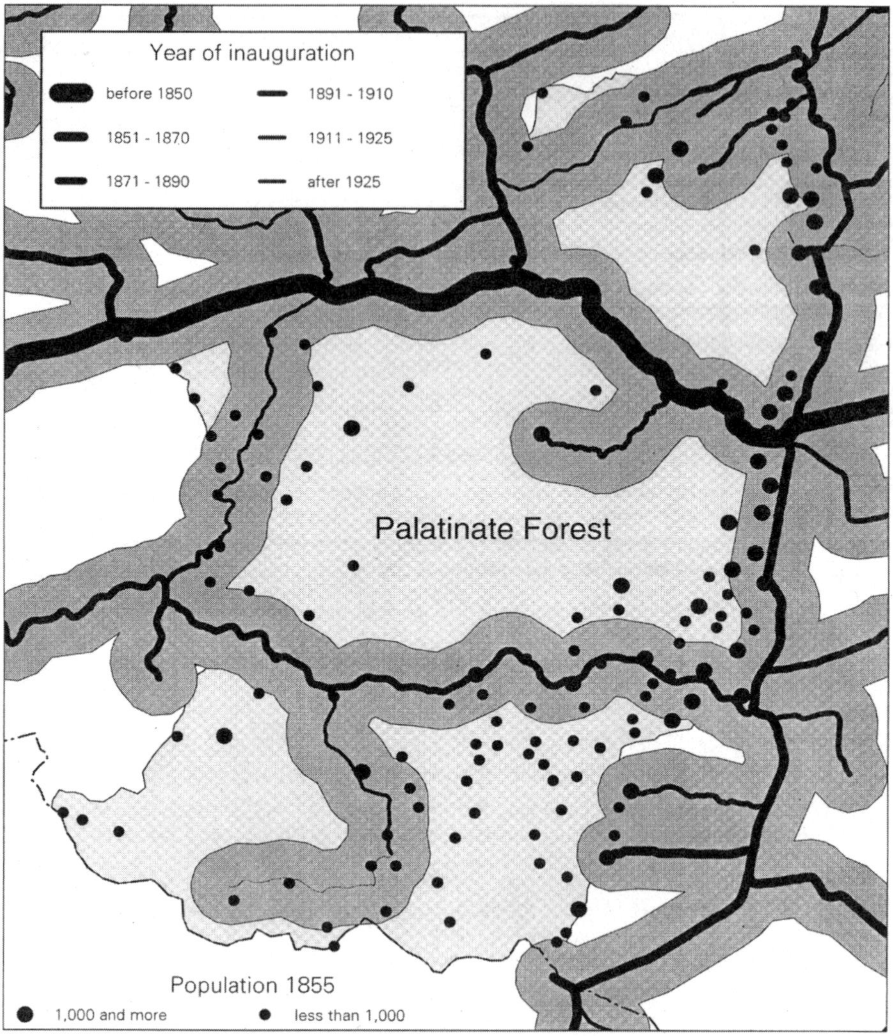

Fig. 7.3 19th century transport network analysis in the Palatinate Forest

In some cases it is not easy to come up with a spatial framework to deal with historical statistics if different and non hierarchical divisions are applied and no constant administrative units over time are available to aggregate top level divisions from that. This was the case with the Great Britain Historical GIS (Gregory & Gilham 1998, Gregory & Southall 1998). In England and Wales different levels of administrative units above the approximately 15,000 parishes were in use (see figure 7.2). In the case of the Great Britain Historical GIS several hundred up to several thousand changes of administrative boundaries took place and have to be stored in a GIS. Gregory & Gilham (1998:10) detected more than 20,000 changes to parish boundaries alone, which were identified and still are being digitised from suitable historical topographic maps. For the detection of changes (see figure 7.2) an approach similar to the pseudo-object GIS approach presented in chapter 4.4 was introduced, using line segments (Arcs) within Arc/Info. [46]

The approach of a historical administrative GIS as it is administered by the Great Britain Historical GIS project extends the goal of an exclusively on census data based approach towards more versatility in terms of continuous coverage of changes, diversity of administrative units covered, and the historical data used. A comparable objective is applied by the project NLKAART (Boonstra 1994).

7.2 Studying historical transport networks

The development of railway networks across Europe and North America had a crucial impact on the regional spatial structure in terms of population, economy et cetera (Kunz 1994). The expansion of railway networks may be conceptualised as a diffusion process, where new parts were added successively. New railways can be defined in time by their opening date. Comparable to the point map approach discussed in chapter 4.4 the line segments used to map the railways may be grouped by periods of installation (Swiaczny 1997).[47] In figure 7.3 an example for the Palatinate Forest (Southwestern Germany) is given, based on a list of railways ordered according to their inauguration date and topographic maps representing the widest spread of railway lines just before World War II. Using the analysis function of the GIS, buffer zones around the railway lines were created to investigate the impact of the new means of transportation on the peripheral areas of the Palatinate Forest. In addition villages and towns are indicated as point locations. In a next step an analysis was performed selecting all settlements lying inside the buffer zone of a certain railway segment. Finally, the population development was analysed differentiated according to the time before and after the opening of railway lines, and settlements close to or far from a railway station.

Again the census data had to be consolidated to time series. Because there was no systematic coverage of administrative changes at hand, the gazetteer method introduced in chapter 4.4 was used to aggregate the old census data to the territorial state of modern communities. As a result of the research it was possible to prove that the expansion of the railway network was not the main cause for a positive development of the population in the Palatinate Forest.

---

[46] See the related website at [http://www.geog.port.ac.uk/gbhgis/]

[47] Swiaczny F (1997) Die Verkehrserschließung des Pfälzerwaldes und ihre Auswirkungen auf die räumlich differenzierte Bevölkerungsentwicklung. In: Anhuf D, Jentsch C (eds) Beiträge zur Landeskunde Südwestdeutschlands und angewandten Geographie (Mannheimer Geographische Arbeiten 46). Geographisches Institut der Universität Mannheim, Mannheim, pp 121-144

7.3 Exploring cultural landscape change[48]

Man and biosphere are closely interrelated with each other. The environment is continuously changing through time ever since the existence of our planet. The impact of mankind on the environment may be understood as a process of adaptation to the living conditions always under change. In the last centuries an increasing part of the world was modified by economic and social activities of human society. Thus, a large portion of the world's surface is now manmade environment and transformed into a cultural landscape.

The close relationship between man and environment is of great interest for geographers and other scientists. The cultural landscape and its elements are observed and stored in inventories, e.g. used for landscape conservation or regional planning. Today, such inventories are usually stored in a GIS. Figure 7.4 lists the sources used for this purpose from a German perspective (Swiaczny & Ott 1999; Ott & Swiaczny 2000). From the great variety of geographical and thematic sources several methodological problems evolve. At this point, two aspects are discussed in greater detail:

- The first issue related to landscape inventories is the conceptualisation of spatial entities and the selection of corresponding spatial object to be stored in a GIS. Besides topographic and other thematic maps, also raster data e.g. from satellite images and digital terrain models are used to model the cultural landscape in a GIS (Lutze & Schultz 1995). The related conceptual problems were discussed in chapter 2.3.

---

[48] Specific references cites in this chapter:

Bauernschmitt G. (1997) Biosphärenreservat Rhön-Konzept für Schutz, Pflege und Entwicklung. In: Ott E. (ed) Zukunft der Kulturlandschaften. Aufgaben und Konzepte nachhaltiger regionaler Entwicklungen. (IKU-Reihe "Kommune und Umwelt" 5) Institut für Kommunale Wirtschaft und Umweltplanung Darmstadt, Frankfurt/M., pp 96-109

Bieler J, Bastian O (1994) Übersicht über wichtige Datenquellen. In: Bastian O Schreiber KF (ed.): Analyse und ökologische Bewertung der Landschaft. G. Fischer, Stuttgart, pp 416-443

Denecke D (1972) Die historisch-geographische Landesaufnahme. Aufgaben, Methoden und Ergebnisse, dargestellt am Beispiel des mittleren und südlichen Leineberglandes. In: Hövermann J, Oberbeck O (ed) Hans-Poser Festschrift. (Göttinger Geographische Abhandlungen 60), Geographisches Institut der Universität Göttingen, Göttingen, pp 401-437

Ewald KC (1996) Traditionelle Kulturlandschaften. Elemente und Bedeutung. In: Konold W (ed): Naturlandschaft-Kulturlandschaft. Die Veränderung der Landschaften nach der Nutzbarmachung durch den Menschen. Ecomed, Landsberg, pp 99-119

Jäger H (1987) Entwicklungsprobleme europäischer Kulturlandschaften. Eine Einführung. Wissenschaftliche Buchgesellschaft, Darmstadt

Kleefeld KD (1994) Historisch-geographische Landesaufnahme und Darstellung der Kulturlandschaftsgenese des Zukünftigen Braunkohlenabbaugebietes Garzweiler II. dissertation, Bonn

Meynen H (1997) Angewandte Historische Geographie in der Stadt. Kulturlandschaftsforschung am Beispiel Köln. In: Kleefeld KD, Burggraaff P (ed) Perspektiven der Historischen Geographie. Seminars für Historische Geographie der Universität Bonn, Bonn, pp 277-283.

- The second aspect is due to the great variety of sources to be used in one system. Their integration brings about several serious problems for the GIS based creation of such an inventory. The spatio-temporal reliability of sources with different scales, different levels of resolution and accuracy or different projections imposes high demands on the expertise of the GIS user.

The inventory of a cultural landscape change is in general not meant as an end in itself. It may be applied for instance to determine the impact of alternative sites or planning scenarios for a development project on the cultural landscape (Czeranaka 1996). Often not only the current state stored in an inventory or the simulation of possible future states are of interest, but the analysis of historical changes of the cultural landscape and the interdependences between man and environment are the focus of the scientific interest. Thus, spatio-temporal processes must be analysed within the GIS. The complexity of the GIS will increase if social and economic processes have to be adopted to the inventory of the physical sphere.

First of all, temporal information is inherent in the entities currently present in the cultural landscape. Often such temporal information is not explicitly given and must be reconstructed from different sources. On the other hand it is necessary to reconstruct historical states of the cultural landscape. The past in present perspective presented in chapter 4.4 might be used to obtain primary information. However, these sources are obviously not complete and have to be amended by historic maps and materials. In practice, the analysis of cultural landscape change has to deal with a great amount of uncertainty and a constant process of re-evaluation and editing of the already available information in the light of evolving new sources (Albrecht 1995).

As a consequence spatio-temporal GIS which are used for the analysis of cultural landscape changes should at least be able to cover transaction time as well as valid time and furthermore different versions of identical objects. As new spatio-temporal objects have to be added at the time they occur during the research process (not in a chronological order), the GIS must be able to integrate updates without effort. A logical data structure based on an Arc-Node can be recommended.

Finally, the architecture of the GIS itself is of interest. As a result of their constraints, layer oriented GIS do not meet all requirements set up by landscape analysis. Pseudo-object and object GIS are preferred used for this purpose because of their ability to deal with different sources and types of spatio-temporal data on one consistent level.

Spatial Entity	Source	Scale	Recording/Updates	GIS data type
Group 1: natural environment				
terrain	digital elevation model	50 x 50 m of raster width	once	3D-model (x,y,z)
slope inclination	derived from DEM	like DEM	once	raster
exposure	derived from DEM	like DEM	once	raster
geology	Geological maps	1:25.000 - 1:200.000	once	polygon
hydrology	topographic and hydrologic maps	1:25.000 - 1:50.000	once to regularly (time sections)	line and polygon
vegetation	vegetation maps, maps of the potential natural vegetation	1:25.000 - 1:200.000	once to regularly (time sections)	polygon
soil types	soil maps	1:5.000 - 1:200.000	once to regularly (time sections)	polygon
soil values	Reichsbodenschätzung	1:500 - 1:2730	Once	polygon
natural regions	"Handbuch der naturräumlichen Gliederung" and geographical land survey 1:200.000	1:100.000 - 1:200.000	Once	polygon
Groups 2/3: cultural landscape				
land use	maps from the "Urkataster", land register maps and data	1:5.000 - 1:25.000 ~1:5.000	Once	polygon
	forestry administration maps	1:5.000 - 1:10.000	regularly (time sections)	polygon
	topographic maps	1:25.000	regularly (time sections)	polygon
	aerial photographs (analysis)	< 1:25.000	once to regularly (time sections)	raster
	satellite images (analysis)		once to regularly (time sections)	raster
land division	"Urkataster"	~1:5.000	Once	polygon
	agricultural maps	1:500 1:2.500	regularly (time sections)	polygon
	DGK5	1:5.000	regularly (time sections)	polygon
ownership structure	maps from the "Urkataster", land registry maps and data	~1:5.000	Once	polygon
	DGK5, land registry maps and data ("Liegenschaftsbuch", data protection!)	1:5.000	regularly (continuous land register)	polygon
administrative units (socio-economic basic conditions)	topographic maps, municipality and district maps, boundary maps, data from external sources	1:25.000 - 1: 200.000	regularly (time sections), as new editions are frequently published at the date of a modification also continuously	polygon
land use restrictions	development plans, land use plans, regional development plans	< 1:50.000	once (updated if necessary)	line and polygon

territorial development	topographic maps, historical atlases, data from external sources	1:25.000 - 1: 500.000	regular (time sections), as new editions are frequently published at the date of a modification also continuously	polygon
settlement development	"Urkataster", thematic data from land registry	~1:5.000	once	polygon
	DGK5, land registry maps and data ("Liegenschaftsbuch", data protection!)	1:5.000	regular (continuous land register)	polygon
	topographic maps	1:25.000	regular (time sections)	point and region
	"Denkmaltopographie"	1:5.000 - 1: 50.000	once (updated if necessary)	point and region
transport network	topographic maps	1:25.000 - 1:100.000	regular (time sections)	line
	historical atlases, external sources	1:100.000 - 1: 500.000	once to continuous	line
abandoned settlements	archaeological land survey, aerial photographs, topographic maps	1:500 - 1:25.000	once (updated if necessary)	point and polygon
evidence of former economic activity	DGK5, land register maps and data	~1:5.000	once	point and polygon
	DGK5, land register maps and data ("Liegenschaftsbuch", data protection!)	1:5.000	regular (continuous land register)	polygon
	topographic maps	1:25.000	regular (time sections)	point (polygon)
evidence of former land use	aerial photographs, temporal classification after external sources	< 1:25.000	once	point, line and polygon
	"Urkataster"/DGK5, temporal classification after external sources	1:5.000	once	point, line and polygon
	topographic map, temporal classification after external sources	1:25.000	once	point, line and polygon

Fig. 7.4 Sources for the analysis of cultural landscape change (own design based on Denecke 1972, Jäger 1987, Bieler and Bastian 1994, Kleefeld 1994, Hake and Grünreich 1994: 400-413, Quasten and Wagner 1996, Bauernschmitt 1997, Ewald 1997, Meynen 1997)

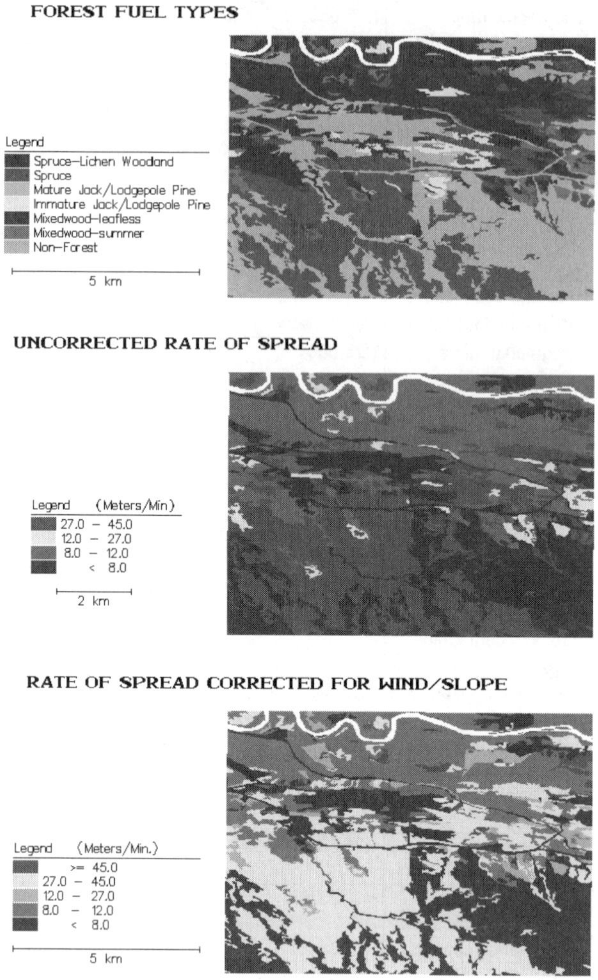

Fig. 7.5 Predicting the rate of spread of forest fires

7.4 Working with cadastres

Cadastral maps are used all around the world to store information on public and private property of land (Dale 1991). Usually physically demarcated in the real world, the cadastral maps show the borders of land parcels, corresponding with legal records covering the ownership of such a parcel. The example of cadastre was already used in chapter 3.4 to explain the concept of versioning. Cadastral maps are commonly maintained by the cadastral offices and provide the highest standard of spatial as well as of temporal accuracy, often using sophisticated GIS technology (Frank 1996). Current cadastral maps and the related legal records are in general not available to the public. Nevertheless old cadastral maps may serve as a valuable source to study the socio-economic structure of a municipality. They are often also the most accurate maps for ancient periods and therefore important sources to help geo-referencing old inaccurate topographical or thematic maps (Vrana 1989).

Taken together, cadastral maps and legal records provide a complete coverage of the development of the property rights as well as of the built environment. Able to track all changes continuously and based on maps with a unique projection, scale and accuracy, cadastral maps are the sources best suited to implement the least common geometry and pseudo-object approaches from chapter 4.4. However, the large scale and the vast number of changes stored in the legal records prevents one from using cadastral maps for an extended area and/or time span.

7.5 Predicting wildfires

A time-integrative GIS can help in wildfire management, a monumental and vital task, as the summer of 2000 has painfully proven. Fire prediction may help to save millions of dollars in natural and manmade resources. Fire prediction is where the real potential lies: Given a current burning fire, where will it likely move next? What combative resources will likely be necessary? What effect does soil moisture or type have on a fire's propagation?

The following example of a raster based analysis is derived from a demo application distributed in the late 1980s by TYDAC Research Inc. (now a division of PCI Geomatics). The demo shows how the company's GIS package SPANS could be used to solve predictive equations to estimate forest fire rate of spread.

The model was based on forest cover groups, a function combined of tree species composition, tree height, and tree density (see figure 7.5). The forest cover groups were reclassified into forest fuel types, which include a measure of stand composition and structure. The rate of fire spread is influenced by weather conditions such as wind speed, humidity, rainfall and temperature. The fuel map and local weather observations were combined in the model in order to estimate the rate of spread (measured in meters per minute). As the rate of spread is also affected by topographic features such as slope and aspect, topographic features were derived from a digital terrain model. The resulting map enables fire fighters to combat wildfires more efficiently and safely.

7.6 Mapping urban growth

Land-use change through time has been a popular research subject for city, county and regional planners. Such research usually involves the use of land-use maps, aerial photographs and/or satellite images. Researchers compare land-use patterns from one time period with another, based on available reference materials.

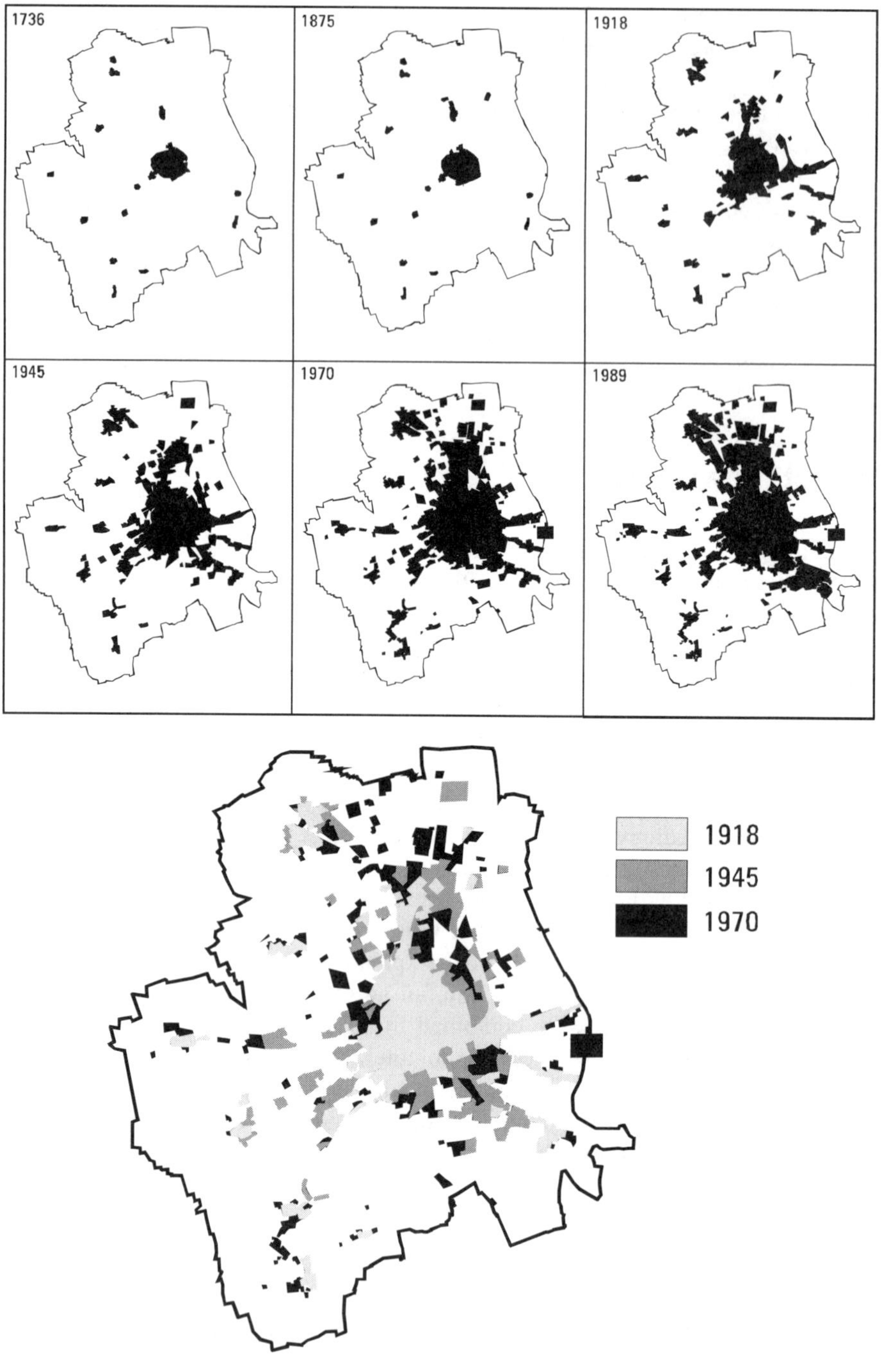

Fig. 7.6 Analysing land use change in the city of Erfurt

Similarly, global change studies compare NASA satellite imagery from the early Landsat program with modern images to monitor the size and composition of tropical rainforests. Thus, statements about the number of square kilometres of forests lost to human development are derived.
Most metropolitan areas face the growing problems of urban sprawl, which include a loss of natural vegetation and open spaces, and a general decline in the spatial extent and connectivity of wildlife habitat and agricultural lands. Residential and commercial development is replacing undeveloped land at an unprecedented rate. Urbanization exerts heavy social, ecological, environmental and climatic pressures on surrounding lands in comparison to its spatial extent.

Temporal urban mapping develops a historical perspective of urban development by combining a variety of data sources into an integrated, multi-scale, and multi-resolution database. Many regional planning organizations or research institutions such as the United States Geological Survey (USGS)[49] utilize such combinations of historic maps and satellite images in conjunction with digital elevation models and local land use maps to analyse and visualize temporal changes in land use. Visual analysis of the database reveals the development of spatial patterns over time. Transportation data layer clearly demonstrate the influence that the transportation infrastructure (roads, railroads) exert on population development.

The example presented in figure 7.6 shows a sequence of six time slices of the built-up area of the city of Erfurt.[50] Applying a sequence of time slice overlays and intersections (see chapter 4.4.5) numerous questions may be answered, for instance the loss of agricultural land in a period of time, or the growth rate of residential or commercial land uses. If a sufficient number of time slices is available, the maps may be transformed into an animation (see chapter 6.2).

7.7 Reconstructing archaeological excavations

In archaeology the mapping of excavations is of great concern. GIS can improve the storage of archaeological data during the field work and the analysis later on. The area of the excavation is usually either triangulated ore measured using arial photography, terrestrial photogrammetry or a GPS receiver. Each object found can be located either directly using x,y,z-coordinates or x,y-coordinates related to a layer representing a certain level of the excavation site. The analysis requires a precise determination of the objects, e.g. in terms of origin, age et cetera. In Figure 7.7 a three-dimensional space is shown in which the artefacts are located on three elevation levels. For each elevation level reached during the excavation a snapshot was created to exactly map all pieces found. In addition, the objects in the example are temporally determined, each object carries a temporal meaning. In the least complex case, all items found are of the same nature and belong to one entity

[49] Cf. http://www.ncgia.ucsb.edu/projects/gig/ncgia.html for further information.

[50] The example is taken from Ott T (1997) Erfurt im Transformationsprozeß der Städte in den neuen Ländern (Erfurter Geographische Studien 6). Selbstverlag, Erfurt, p 124

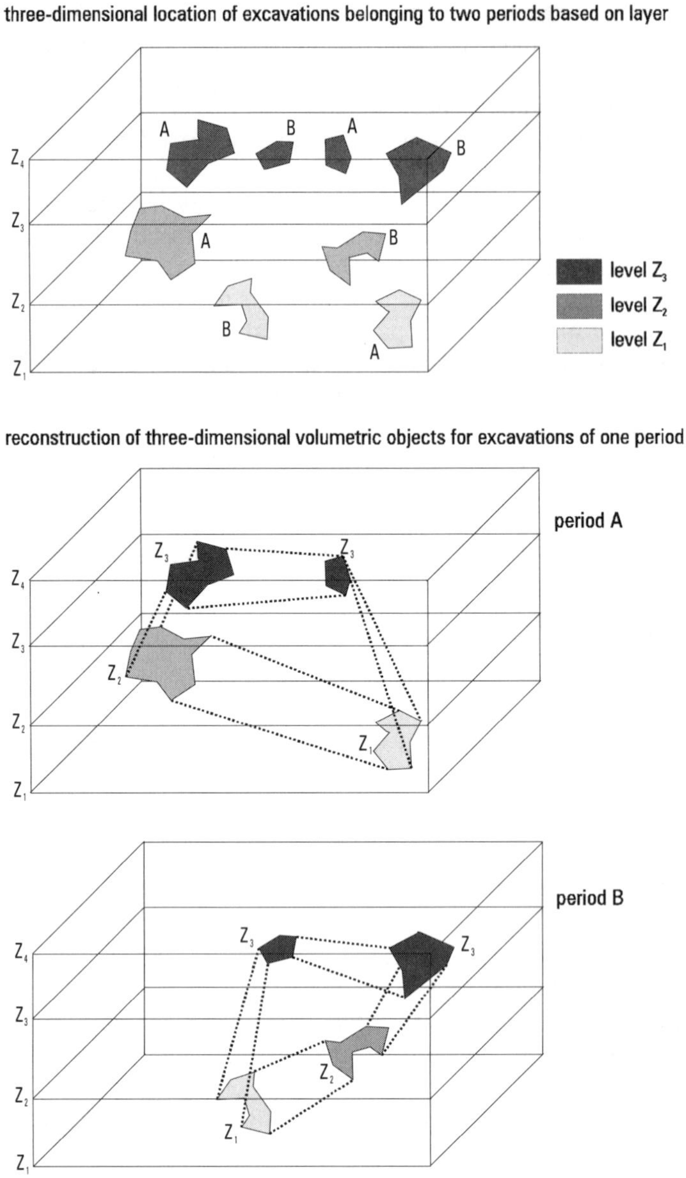

Fig. 7.7 Locating archaeological artefacts

which has to be reconstructed using the evidence drawn from the excavated artefacts. In a first step from the three-dimensional data set depicted in figure 7.7 several maps may be drawn. Each of them can show either the location of objects differentiated according to age or the elevation level were they were found. In this case a time slice approach is transferred to cover elevation and temporal data is assigned as attribute data, mapped similar to the examples given for temporal point maps in chapter 4. For analytical purposes the objects might be treated as volumetric objects. Their distribution within a three-dimensional sphere can be visualized and underlying structure may be explored.

7.8 Mining for business data

Business applications profit from the complex analytical and mapping possibilities provided by common GIS packages. In recent years business applications of GIS represent the fasted growing market for Desktop GIS, exceeding all other fields where GIS are traditionally used for a long time. The analytical functions of a GIS are often used to detect changes and insofar business GIS in general are time-integrative GIS. The use of GIS aim to analyse, where e.g. the development of the market share of a company lags behind the competition. On the other hand prognostic models are applied, for instance to identify evolving markets from the analysis of past growth.

Often the spatial reference system of business data is diverse or changes through time. As a consequence, the valuable time series data captured cannot be processed for a period long enough to build time series. Or they cannot be compared with data from other sources. The completion of time series data, as already described as a method in chapter 4, is of great importance. In practice the redesign of market areas may impose a new perspective on the historic data stock. A typical application could be to determine the past development of sales aggregated to the changed spatial divisions.

During the last years "data warehousing" is a hot topic in business computing. Data warehouses are used to access the hidden dimension of strategic information enclosed in operational data. To make use of the potential of information available in different branches of a company, the data stored as part of everyday work have to selected, consolidated and aggregated. Not every single information necessary to deal with day to day requests needs to be stored in the data warehouse. Other data have to be processed or aggregated first, before it is of use for analytical or prognostic purposes. Nevertheless, all data stored have to be accompanied by metadata which allows to access the content or meaning of datasets. In the end, the data warehouse provides the user with pre-structured data and tools to access and explore the information content. The aggregation of time series data and the continual update of such time series from the operational processes is an essential task of a data warehouse. Even under changing circumstances like new spatial divisions the data warehouse must be able to provide the user with the time series data he needs. Therefore GIS with temporal abilities are frequently integrated in the data warehouse, both as a means of data collection, processing and storage as well as a tool for the exploration of spatio-temporal trends (see figure 7.8).

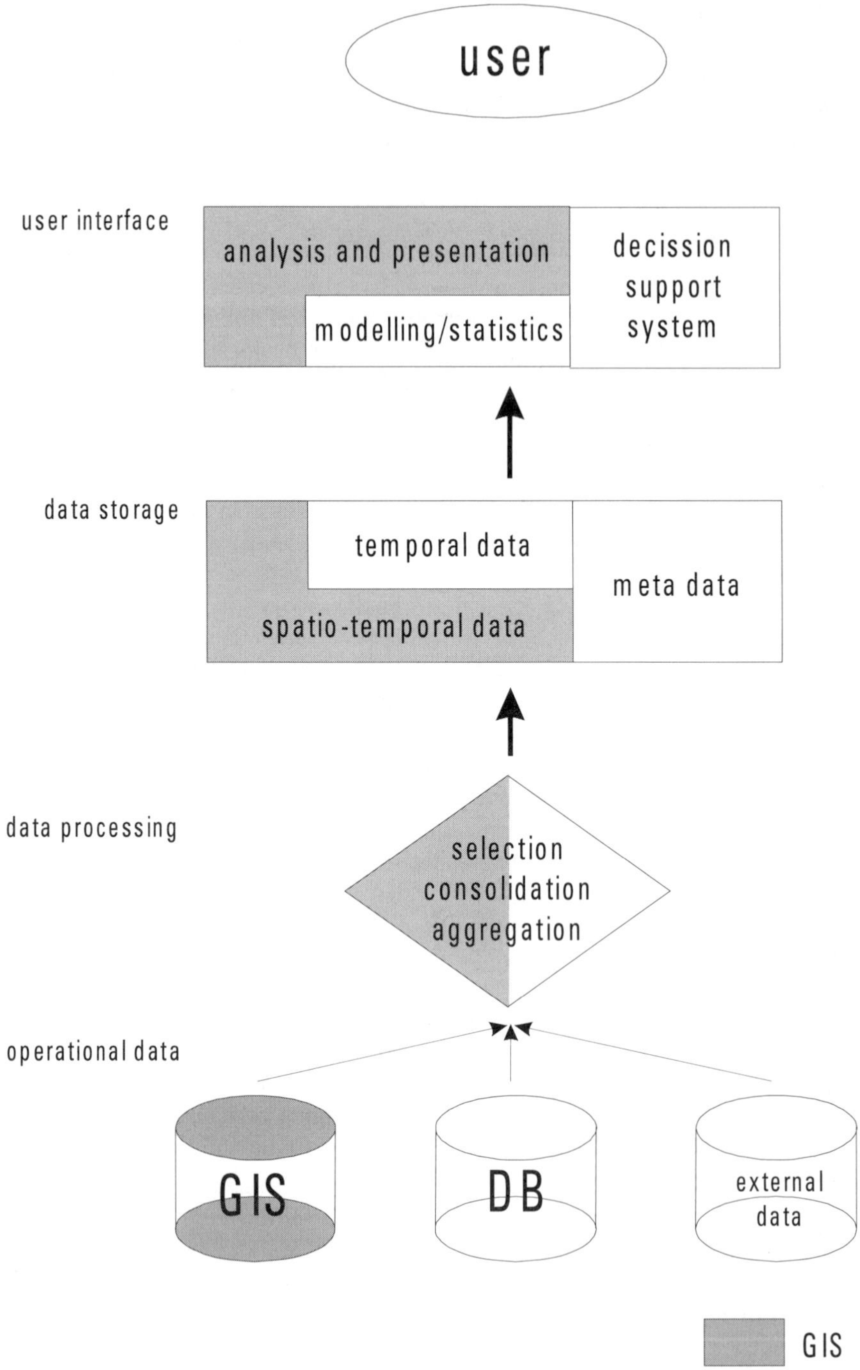

Fig. 7.8 Structure of a data warehouse

8 Conclusion

> "From henceforth, space by itself, and time by itself, have vanished into the merest shadows and only a kind of blend of the two exists in its own right."
> (Hermann Minkowski)[51]

8.1 Summary

Main purpose of a temporal or time-integrative GIS is, to reproduce temporal processes or sequences of events of the real world in a model in such a way, to make them accessible for spatial query, analysis and visualization. In order to be able to represent and analyse the temporal and spatial interrelations, the modelling of spatio-temporal data requires extraordinary assiduousness. An important prerequisite for the successful GIS use is therefore a systematic plan of action for the modelling of the significant questions and the gradual steps of abstraction from the conceptual model of the real world up to the physical data model in the computer. One must consider, which entities and which relations between these entities have to be represented in a model of the real world and which analysis approaches should be possible with this model.

The specific problems of a time-integrative GIS are based on the necessity of simultaneous spatial and temporal reliability. If the individual information layers of the GIS are not to be explored independently, it is of fundamental importance, that on the level of the physical data model, identical points (even as element of a line or polygon) are represented by exactly one pair of coordinates. This holds true for identical points in different thematic layers as well as identical points in different temporal layers or different scales. The demand for spatial and temporal reliability leads to the most prominent problem with the construction of a historical GIS: the representation of temporal information in a conceptual data model and the reproduction of temporal data in a logical data model. The structured proceeding described in this book is not to be understood as an end in itself. It rather permits a first overview of the system to be analysed. The more exact the elements and interactions of the system are studied and transformed into a conceptual model against the background of theoretical knowledge, the easier and better it will be possible to form the logical entity-relationship model in the next step. Another

---

[51] Quoted in Newman JR (1956) The world of mathematics. Simon & Schuster, New York

advantage of this procedure results from the fact that missing links and relations can already be uncovered, before the actual data acquisition has started. Since the construction of the data objects exerts strong influence on the possibilities of analysis, their feasibility can already be checked at an early stage and modified if necessary. Furthermore the entity-relationship model forms a basis of decision for the selection of data sources, as it permits to identify inappropriate or redundant information and thus may save operating time and costs. However, the significance of these theoretical considerations is qualified by the characteristics of the available sources, since these are not available or accessible in every desired temporal and/or spatial resolution.

8.2 Outlook

"Much remains to be done before a true temporal GIS can be realised. As was the case in current spatial database systems, the introduction of representational power and analytical capabilities in commercial systems will itself be a temporal process" (Peuquet 1999:101). Although the basic computer technology for spatio-temporal analysis exists, the GIS community must undergo a paradigm shift to fully appreciate the benefits of time-integrative GIS. It is not just a matter of collecting time-based data within a GIS, but also developing a new way of thinking about time in a spatial sense. It was Albert Einstein who first used the words "space" and "time" together in the same sentence when he developed his theory of general relativity. Oddly enough, the concept of a time-integrative GIS is remarkably similar to work being done by modern physicists who believe in a "unified theory": that one set of physical rules defines space, time and matter in one neat package. Perhaps an earthly application must be developed and implemented that demands new ways of thinking about spatio-temporal analysis before the GIS software industry can catch up to the science (Castagneri 1998).

In order to create a truly time-integrative GIS the focus has to change from spatial to temporal and from analysing changes between events to the analysis of change itself. However, the methodological challenges to reach this objective will even increase in the future. This means that only a small group of experts in the field of methodology, database design and GIS will be able to promote the future development. With an increasing complexity of advanced methodological and technical solutions for spatio-temporal GIS applications it seems more and more unlikely that such technology will be made available to the community of average users in standard GIS packages. On the other hand, if the professional GIS developers came up with a fancy TGIS extension which would hide the methodological problems behind some simple dialogue boxes, the potential pitfalls and drawbacks would by far outnumber the additional analytic possibilities offered.

What in the context of this book should be reached is a deeper understanding of the conceptualisation and implementation problems of time into available GIS packages. As stated above, spatio-temporal GIS increase the complexity of methodological problems often beyond the qualifications and resources of average GIS users. GIS experts on the other hand frequently lack knowledge in the respective field of application. From the authors' point of view a closer cooperation between the GIS world and the experts in the field of spatio-temporal GIS on the one side, and the end user on the other side could help to solve many problems using present-day technology and methods.

It's about time to start this cooperation right away, and we hope that this book can contribute a little to find solutions a wide variety of potential users are looking for. We hope to encourage others to share their expertise and amend to the content of this book, or to fix shortcomings which we are aware will appear during the work with the book. So, please feel invited to visit our website at http://www.time-gis.de and share your thoughts and comments on the book with us and others interested in time-integrative GIS.

9 Appendix

9.1 Research bibliography

Aamodt G (2000) A statistical model to analyse spatio-temporal data. In: Fotheringham AS, Wegener M (eds) Spatial models and GIS (GISData 7). Taylor & Francis, London, pp 235-244

Abbott MB (1991) Hydroinformatics: information technology and the aquatic environment. Avebury Technical, Aldershot

Abdelmoty AI, Williams MH, Patron NW (1992) Deduction and deductive databases for geographic data handling. In: Goodchild M, Gopal S (eds) The accuracy of spatial databases. Taylor & Francis, London, pp 443-464

Abiteboul S, Herr L, Van den Bussche J (1995) Temporal connectives versus explicit timestamps in temporal query languages. In: Clifford J, Tuzhilin A (eds) Recent advances in temporal databases. Proceedings of the International Workshop on Temporal Databases, Zürich, Switzerland, 17-18 September 1995 (Workshops in Computing). Springer, Berlin/Heidelberg/New York, pp 43-47

Abler R (1993) GPS, GIS and geography in the 1990s. Professional Geography 45:131-139

Adam NR, Gangopadhyay A (1997) Database issues in Geographic Information Systems (Kluwer International Series on Advances in Database Systems 6). Kluwer, Boston

Agrawal R, Srikant R (1995) Mining sequential patterns. Proceedings of the 11th International Conference on Data Engineering, pp 3-14

Ahn I, Snodgrass R (1986) Performance evaluation of a temporal database management system. In: Zaniolo C (ed) Proceedings of the 1986 ACM SIGMOD International Conference on Management of Data, Washington, DC, May 28-30, 1986, pp 96-107

Alagic S (1997) A temporal constraint system for object-oriented databases. Proceedings of the Second International Workshop on Constraint Database Systems (LNCS 1191), pp 208-218

Albrecht J (1994) Methoden und Modelle in Geographischen Informationssystemen. In: Aurada KD (ed) Beiträge des 10. Kolloquiums für Theorie und quantitative Methoden in der Geographie (Greifswalder Geographische Arbeiten 11). Universität Greifswald, Institut für Geographie, Greifswald, pp 104-113

Albrecht J (1995) A framework for dealing with spatially and temporally undetermined geographic objects. In: Kremers H, Pillmann W (eds) Raum und Zeit in Umweltinformationssystemen. 9th International Symposium on Computer Science for Environmental Protection CSEP '95 (Umwelt-Informatik aktuell). Metropolis-Verlag, Marburg, pp 143-150

Aldenderfer MS, Maschner HDG (eds) (1996) Anthropology, space, and Geographic Information Systems (Spatial Information Series). Oxford University Press, New York

Alexandroff PS (1961) Elementary concepts of topology. Dover Publications, New York

Alfarmanesh H, McLeod D, Knapp D, Parker A (1985) An existensible object-oriented approach to databases for vlsi/cad. Proceedings of the 6th International Conference on Very Large Databases. IEEE, New York, pp 86-91

Allen E, Edwards G, Bédard Y (1995) Qualitative causal modeling in temporal GIS. In: Frank AU, Kuhn W (eds): Spatial information theory-a theoretical basis for GIS. Proceedings COSIT '95 (LNCS 988). Springer, Berlin/Heidelberg/New York, pp 397-412

Allen J (1983) Maintaining knowledge about temporal intervals. Communications of the ACM 26:832-843

Allen JF, Hayes PJ (1985) A common-sense theory of time. In: Aravind KJ (ed) Proceedings of the 9th International Joint Conference on Artificial Intelligence, Los Angeles, Morgan Kaufmann, San Mateo, pp 528-531

Allen KMS, Green SW, Zubrow EBW (eds) (1990) Interpreting space: GIS and archaeology. Taylor & Francis, London

Al-Taha KK (1992) Temporal reasoning in cadastral systems. Unpublished Ph.D. thesis, University of Maine, Orono

Al-Taha KK, Barrera R (1990a) Temporal data and GIS: an overview. Proceedings of GIS/LIS '90, Anaheim. ACSM, Bethesda, pp 244-254

Al-Taha KK, Barrera R (1990b) Temporal databases: their applicability to the ISDS. University of Maine, Orono

Al-Taha KK, Frank A (1991) Temporal GIS keeps data current. GIS World (ed) International GIS Sourcebook. GIS World, Fort Collins, pp 382-388

Al-Taha KK, Snodgrass RT, Soo MD (1993) Bibliography on spatiotemporal databases. SIGMOD Record 22:59-67

Al-Taha KK, Snodgrass RT, Soo MD (1994) Bibliography on spatiotemporal databases. International Journal of Geographical Information Systems 8:95-103

American National Standards Institute (ANSI) (1997a) Spatial Data Transfer Standard (SDTS). Logical specifications. ANSI, New York

American National Standards Institute (ANSI) (1997b) Spatial Data Transfer Standard (SDTS). Features. ANSI, New York

Ameskamp M, Clauss M, Heinrich U (1995) Developing an interactive spatial query interface for ARC/INFO Map Collections. In: Kremers H, Pillmann W (eds) Raum und Zeit in Umweltinformationssystemen. 9th International Symposium on Computer Science for Environmental Protection CSEP '95 (Umwelt-Informatik aktuell). Metropolis-Verlag, Marburg, pp 168-174

Andersen M (1992) Spatial analysis of two-species interactions. Oecologia 91:134-140

Andler S, Hansson J, Eriksson J, Mellin J, Berndtsson M, Eftring B (1996) DeeDS towards a distributed and active real-time database system. SIGMOD Record 25:38-40

André E, Rist T (1996) Coping with temporal constraints in multimedia presentation planning. National Conference on Artificial Intelligence and Eighth Innovative Applications of Artificial Intelligence Conference, AAAI 96/IAAI 96, August 4-8, 1996, Portland, OR, pp 142-147

Androutsopoulos I, Ritchie GD, Thanisch P (1995) Experience using TSQL2 in a natural language interface. In: Clifford J, Tuzhilin A (eds) Recent advances in temporal databases. Proceedings of the International Workshop on Temporal Databases, Zürich, Switzerland, 17-18 September 1995 (Workshops in Computing), pp 113-132

Antenucci JC, Brown K, Croswell P, Kevany M, Archer H (1991) Geographic Information Systems: a guide to the technology. Chapman & Hall, New York

Aranha RFM, Ganti V, Narayanan S, Muthukrishnan CR, Prasad STS, Ramamritham K (1996) Implementation of a real-time database system. Information Systems 21:55-74

Arbia G (1989) Spatial data configurations in statistical analysis of regional economic and related problems (Advanced Studies in Theoretical and Applied Econometrics). Kluwer Academic Publishers, Dordrecht

Arctur D, Sargent P (1998) Introduction to object-oriented GIS technology [http://www.sargents.demon.co.uk/Philip/oo+gis/index.htm]

Ariav G (1986) A temporally oriented data model. ACM Transactions on Database Systems 11:499-527

Aritsugi M, Tagashira T, Amagasa T, Kanamori Y (1997) An approach to spatio-temporal queries. Interval-based contents representation of images. In: Hameurlain A, Tjoa AM (eds) Database and expert systems applications, Proceedings of the 8th International Conference, DEXA '97, Toulouse, September 1-5, 1997 (LNCS 1308). Springer, Berlin/Heidelberg/New York, pp 202-213

Arlinghaus SL (ed) (1994) Practical handbook of digital mapping: terms and concepts. CRC Press, Boca Raton

Arlinghaus SL (ed) (1996) Practical handbook of spatial statistics. CRC Press, Boca Raton

Armenakis C (1991) The temporal dimension of geographical data. Technical Report, Applied Research and Technology Service, Geographical Services Division. Energy, Mines and Resources Canada, Ottawa

Armenakis C (1992a) Electronic mapping of time-dependent data. Proceedings of ASPRS/ACSM/RT, Washington, DC. ACSM, Bethesda, pp 445-454

Armenakis C (1992b) Estimation and organization of spatio-temporal data. Paper presented at the Canadian Conference on GIS '92

Armenakis C (1993) Map animation and hypermedia: tools for understanding changes in spatio-temporal data. Proceedings of the Canadian Conference on GIS '93, Ottawa, Canada. The Canadian Institute of Geomatics, Ottawa, pp 859-868

Armstrong M (1988) Temporality in spatial databases. Proceedings of GIS/LIS '88, San Antonio, pp 880-889

Aronoff S (1989) Geographic Information Systems: a management perspective. WDL Publisher, Ottawa

Asch K (1999) GIS in Geowissenschaften und Umwelt. Springer, Berlin/Heidelberg/New York

Aspinall RJ, Pearson DM (1995) Describing and managing uncertainty of categorical maps in GIS. In: Fisher P (ed) Innovations in GIS 2. Taylor & Francis, London, pp 71-83

Atkinson M et al. (eds) (1992) The object-oriented database system manifesto. In: Bancilhon F, Delobel C, Kanellakis P (eds) Building an object-oriented database system: The story of O2. Kaufmann, San Mateo, pp 3-20

Atkinson PM, Tate NJ (eds) (1999) Advances in remote sensing and GIS analysis. John Wiley & Sons, Chichester/Weinheim

Augustin NH, Mugglestone M, Buckland S (1996) An autologistic model for the spatial distribution of wildlife. Journal of Applied Ecology 33:339-347

Ayer AJ (1984) The central questions of philosophy. Penguin, London

Bachmann P (1993) Handbuch der Satellitennavigation. GPS, Technik-Geräte-Anwendung. Motorbuch-Verlag, Stuttgart

Bagai R, Orgun MA (1995) A temporal paraconsistent relational algebra for incomplete and inconsistent information. Proceedings of the 33rd Annual ACM Southeast Conference, Clemson, pp 240-248

Bagg J (1996) Don't know where, don't know when-using imprecise information. Paper presented to the 1996 Conference of the UK Association for History and Computing. Queen Mary & Westfield College, London

Bagg J, Ryan N (1996) Modelling historical change in Southern Corsica: temporal GIS development using an extensible database system. Proceedings of the GIS Research UK 1996 Conference, Canterbury, pp 75-77

Bagg J, Ryan N (1997) Modelling historical change in Southern Corsica: temporal GIS development using an extensible database system. In: Kemp Z (ed) Innovations in GIS 4. Taylor & Francis, London, pp 42-55

Bähr HP, Vögtle T (1999) GIS for environmental monitoring. Schweizerbart, Stuttgart

Bähr U, Singer C, Kiesseling W (1994) Zur Systematik räumlicher Operatoren in Geo-Datenbanken. GIS 4:13-21

Bailey TC (1995) A review of statistical spatial analysis in Geographical Information Systems. In: Fotheringham S, Rogerson P (eds) Spatial analysis and GIS. Taylor & Francis, London, pp 13-44

Bair J, Böhlen M, Jensen CS, Snodgrass RT (1997) Notions of upward compability of temporal query languages. Wirtschaftsinformatik 39:25-34

Balownew O, Bode T, Breunig M (1995) Modellierung und Verwaltung geometrischer Objekte für 3D-UIS. In: Kremers H, Pillmann W (eds) Raum und Zeit in Umweltinformationssystemen. 9th International Symposium on Computer Science for Environmental Protection CSEP '95 (Umwelt-Informatik aktuell). Metropolis-Verlag, Marburg, pp 151-159

Barnert R, Knolmayer G, Myrach T (1995) Rules for ensuring referential integrity in a time-extended relational model. (Working paper 59), University of Bern, Institute of Information Systems, Bern

Barrera R, Frank A, Al-Taha K (1990) Temporal relations in GIS: a workshop. Temporal GIS workshop. University of Maine, Orono

Barrera R, Frank A, Al-Taha K (1991a) Temporal relations in Geographic Information Systems. NCGIA Technical Report 91.4. NCGIA, Santa Barbara

Barrera R, Frank A, Al-Taha K (1991b) Temporal relations in Geographic Information Systems: a workshop at the University of Maine. SIGMOD Record 20:85-91

Bartelme N (1995) Geoinformatik. Modelle-Strukturen-Funktionen. Springer, Berlin/Heidelberg/New York

Baruah A, Bestavros SK (1996) Timely and fault-tolerant data access from broadcast disks. Paper presented at the Fifth International Conference on Information and Knowledge Management, Workshop on databases: active and real-time, Rockyville, MD, November 12-16, 1996

Basoglu U, Morrison J (1978) The efficient hierarchical data structure for the US historical boundary file. In: Dutton G (ed) Harvard Papers on Geographic Information Systems. Addison-Wesley, Reading

Batty P (1992) Exploiting relational database technology in a GIS. Computers & Geosciences 18:453-462

Baudinet M, Chomicki J, Wolpher P (1995) Constraint-generating dependencies. In: Gottlob G, Vardi MY (eds) Database Theory-ICDT'95, Fifth International Conference, Prague, January 11-13, 1995, Proceedings (LNCS 893). Springer, Berlin/Heidelberg/New York, pp 322-337

Bauer M (1997) Vermessung und Ortung mit Satelliten, 4th ed. Wichmann, Heidelberg

Beaumont JR (1992) GIS and market analysis. In: Maguire DJ, Goodchild MF, Rhind DW (eds) Geographical Information Systems. Longman, Harlow, pp 139-151

Beech D, Mahbod B (1988) Generalized version control in an object-oriented database. Proceedings of the Fourth Conference on Data Engineering. IEEE, New York, pp 14-22

Behr FJ (1998) Strategisches GIS-Management. Grundlagen und Schritte zur Systemeinführung. Wichmann, Heidelberg

Bell SBM, Chadwick RA, Cooper APR, Mason DC, O'Connell M, Young PAV (1990) Handling four dimensional geo-code data. In: Brassel KE (ed) Proceedings of the Fourth International Symposium on Spatial Data Handling. Department of Geography, University of Zurich, Zurich, pp 918-927

Beller A (1991) Spatial/temporal events in a GIS. Proceedings of GIS/LIS '91, Atlanta. ACSM, Bethesda, pp 766-775

Beller A, Giblin T, Le K, Litz S, Kittel T, Schmiel D (1991) Temporal GIS prototype for global change research. Proceedings of GIS/LIS '91, Atlanta, pp 752-765

Belward AS, Valenzuela CR (eds) (1991) Remote sensing and Geographical Information Systems for resource management in developing countries (Eurocourses: Remote Sensing 1). Kluwer Academic, Dordrecht

Benatallah B (1996a) Towards a hybrid approach for object-oriented schema evolution management. Proceedings of the Fourth Magrebian Conference on Artificial Intelligence and Software Engineering, pp 97-113

Benatallah B (1996b) Un compromis: modification et versionnement de schéma. Actes des 12ième Journées de Bases de Données Avancées, Cassis, pp 373-393

Benatallah B, Fauvet MC (1997) Le point sur évolution du schéma d'une base d'objects. Revue L'object 3

Bennett RJ (1981) Spatial and temporal analysis: spatial time series. In: Wrigley N, Bennett RJ (eds) Quantitative geography: a British view. Routledge & Kegan, London/Boston/Henley, pp 97-103

Bennett RJ (1989) Whither models in geography in a post-welfare world? In: MacMillan B (ed) Remodelling geography. Basil Blackwell, Oxford, pp 273-290

Berndtsson M, Hansson J (1995) Issues in active real-time databases. Proceedings of the First International Workshop on Active and Real-time Database Aystems ARTDB-95 (Workshops on computing), pp 142-157

Berndtsson M, Hansson J (1996) Workshop report: the First International Workshop on Active and Real-time Database Systems (ARTDB-95). SIGMOD Record 25:64-66

Bernhardsen T (1999) Geographic Information Systems, 2nd ed. John Wiley & Sons, New York/Weinheim

Berrill A, Moon G (1992) An object-oriented approach to an integrated GIS system. Proceedings of GIS/LIS '92, San Jose, pp 59-63

Berry BJL (1964) Approaches to regional analysis: a synthesis. Annals of the Association of American Geographers 54:2-11

Berry JK (1993) Beyond mapping: concepts, algorithms, and issues in GIS. GIS World, Fort Collins

Berry JK (1995) Spatial reasoning for effective GIS. GIS World, Fort Collins

Bertin J (1977) La graphique et le traitement graphique de l'information. Flammarion, Paris

Bertino E, Bettini C, Ferrari E, Samarati P (1996a) A temporal access control mechanism for database systems. IEEE Transactions on Knowledge and Data Engineering 8:67-80

Bertino E, Bettini C, Ferrari E, Samarati P (1996b) Supporting periodic authorizations and temporal reasoning in database access control. In: Vijayaraman TM, Alejandro P; Buchmann C; Mohan N; Sarda L (eds) VLDB '96, Proceedings of 22nd International Conference on Very Large Data Bases, September 3-6, 1996, Bombay, pp 472-483

Bertino E, Bettini C, Ferrari E, Samarati P (1997) Decentralized administration for a temporal access control model. Information Systems 22:223-248

Bertino E, Ferrari E, Guerrini G (1997) T_Chimera: a temporal object-oriented data model. International Journal on Theory and Practice of Object Systems 3:103-125

Bertino E, Ferrari E, Guerrini G (1998) Navigational accesses in a temporal object model. IEEE Transactions on Knowledge and Data Engineering 10:656-665

Bertossi L, Siu B (1996) Answering historical queries in databases. In: Zelkowitz MV, Straub P (eds) Proceedings of the XVI International Conference of the Chilean Computer Science Society, Valdivia, November 13-15, 1996, pp 56-66

Bestavros A (1996) Advances in real-time database systems research. SIGMOD Record 25:3-7

Bestavros A, Braoudakis S (1996) Value-cognizant speculative concurrency control for real-time databases. Information Systems 21:75-101

Bestavros A, Lin KJ, Son S (1996) Report on first international workshop on realtime database systems. SIGMOD Record 25:50-52

Bettini C, Wang XS, Jajodia S (1996a) A general framework and reasoning models for time granularity. In: Chittaro L et al. (eds) Proceedings of the Third International Workshop on Temporal Representation and Reasoning TIME '96. IEEE, Washington, pp 104-111

Bettini C, Wang XS, Jajodia S (1996b) Testing complex temporal relationships involving multiple granularities and its application to data mining. Proceedings of the 15th ACM SIGACT-SIGMOD-SIGART Symposium on Principles of Database Systems, pp 68-78

Bettini C, Wang XS, Jajodia S (1997) Satisfiability of quantitative temporal constrains with multiple granularities. Proceedings of the Third International Conference on Principles and Practice of Constraint Progamming, pp 435-449

Bettini C, Wang XS, Jajodia S, Lin JL (1998) Discovering temporal relationship patterns with multiple granularities in time sequences. IEEE Transaction on Knowledge and Data Engineering 10:222-237

Beven KJ, Moore ID (eds) (1993) Terrain analysis and distributed modelling in hydrology (Advances in Hydrological Processes). John Wiley & Sons, Chichester

Bezzera EP, Schiel U, Ferneda E (1997) Mito-manipulation of incomplete temporal objects. Proceedings of the XII Simposio Brasileiro de Banco de Dados, pp 364-374

Bill R (1991a) Datenbanktechnologie-Überblick und Einsatz in Geo-Informationssytemen. In: Schilcher M (ed) Geo-Informatik-Anwendungen, Erfahrungen, Tendenzen. Siemens-Nixdorf, München, pp 37-52

Bill R (1991b) Zur Eignung moderner Geo-Informationssysteme für Belange der Umweltinformatik. In: Günther O, Kuhn H et al. (eds) Workshop Umweltinformatik 1990. Forschungsinstitut für anwendungsorientierte Wissensverarbeitung an der Universität Ulm FAW-B-91002/II. FAW, Ulm, pp 189-197

Bill R (1992) Raum und Zeit-neue Herausforderungen an Geo-Informationssysteme aus dem Umweltbereich. In: Günther O, Schulz KP, Seggelke J (eds) Umweltanwendungen Geographischer Informationssysteme. C.F. Müller, Karlsruhe, pp 255-264

Bill R (1996) Grundlagen der Geo-Informationssysteme, vol. 2: Analysen, Anwendungen und neue Entwicklungen. Wichmann, Heidelberg

Bill R, Fritsch D (1994) Grundlagen der Geo-Informationssysteme, vol. 1: Hardware, Software und Daten. Wichmann, Heidelberg

Bimbo AD, Vicario E, Zingoni D (1995) Symbolic description and visual querying of image sequences using spatio-temporal logic. IEEE Transactions on Knowledge and Data Engineering 7:609-622

Biondini ME, Mielke PW, Redente EF (1988) Permutation techniques based on euclidian analysis spaces: a new and powerful statistical method for ecological research. Coenoses 3:155-174

Birkin M, Clarke G, Wilson A, Clarke M (1996) Intelligent GIS: location decisions and strategic planning. GeoInformation International, Cambridge

Bishop I (1994) The role of visual realism in communicating and understanding spatial change and process. In: Hearnshaw HM, Unwin DJ (eds) Visualization in GIS. John Wiley & Sons, Chichester, pp 60-64

Bittner T, Frank AU (1997) An introduction to the application of formal theories to GIS. In: Dollinger F, Strobl J (eds) Angewandte Geographische Informationstechnologie IX (Salzburger Geographische Materialien). Institut für Geographie, Universität Salzburg, pp 11-22
Blaschke T (1999) Umweltmonitoring und Umweltmodellierung. Wichmann, Heidelberg
Blobel A, Luttermann H (1995) Forschungsbericht zum Projekt CHRONOS: Integration von Zeit in GIS. Forschungsinstitut für anwendungsorientierte Wissensverarbeitung an der Universität Ulm, FAW-TR-95007 FAW, Ulm [http://www-winfo.uni-siegen.de/winfo/ deutsch/ publikationen/papers/TR95-01.ps.gz]
Bludau-Hary A, Bachmann M, Steinborn W (1995) Erderkundung leichtgemacht. Perthes, Gotha
Bocca J, Hermosilla L (1992) A preliminary study of the performance of MegaLog. Technical Report ECRC-92-34. ECRC, München
Böhlen M, Jensen CS (1996) Seamless integration of time into SQL. Technical Report R-96-2049. University of Aalborg, Department of Computer Science, Aalborg
Böhlen M, Jensen CS, Skjelluag B (1997) Spatio-temporal database support for legacy applications. Technical Report 20, Time Center, Aalborg
Böhlen MH, Jensen CS, Snodgrass RT (1995) Evaluating the completeness of TSQL2. In: Clifford J, Tuzhilin A (eds) Recent advances in temporal databases. Proceedings of the International Workshop on Temporal Databases, Zürich, Switzerland, 17-18 September 1995 (Workshops in Computing), pp 153-172
Böhlen MH, Snodgrass RT, Soo MD (1996) Coalesing in temporal database. Proceedings of the 22nd International Conference on Very Large Databases, pp 180-191
Böhler W, Müller H, Weis N (1999) Bearbeitung historischer Karten mit digitaler Bildverarbeitung. In: Ebeling D (ed) Historisch-thematische Kartographie: Konzepte–Methoden–Anwendungen. Verlag für Regionalgeschichte, Bielefeld, pp 126-136
Bonham-Carter GF (1994) Geographic Information Systems for geoscientists: modelling with GIS (Computer Methods in the Geosciences 13). Pergamon, Kidlington
Boonstra O (1987) NLKAART–een computerprogramma voor het tekenen van een historische kaart van nederlandse gemeenten. Historisch Geografisch Tijdschrift 4:145-147
Boonstra OWA (1994) Mapping the Netherlands, 1830-1994: The use of NLKAART. In: Goerke M (ed) Coordinates for historical maps. A workshop of the association for history and computing. Scripta Mercaturae, St. Katharinen, pp 156-161
Borchert JR (1987) Maps, geography and geographers. The Professional Geographer 39:387-389
Borchert R (1994) ATKIS und seine interdisziplinären Anwendungsmöglichkeiten in anderen Fachinformationssystemen. In: Pfaff-Schley H, Schimmelpfeng L (ed) EDV-Einsatz in Umweltschutz und Landschaftsplanung. Springer, Berlin/Heidelberg/New York, pp 29-41
Borgers A, Witlox F (1997) Modellen voor ruimtelijk gedrag. In: Hendriks P, Ottens H (eds) Geografische informatie systemen in ruimtelijk onderzoek. Van Gorcum, Assen, pp 165-182
Boyd S (1997) Summarizing time-varying data. Proceedings of the Fourteenth National Conference on Artificial Intelligence and Ninth Innovative Applications of Artificial Intelligence Conference, AAAI 97/IAAI 97, July 27-31, 1997, Providence, p 824
Branding H, Buchmann AP (1995) On providing soft and hard real-time capabilities in an active DBMS. Proceedings of the First International Workshop on Active and Real-time Database Systems ARTDB-95 (Workshops in Computing), pp 158-169
Brassel K, Vckovski A, Schmitt E (1998) Informatisierung der Geographie. Geographica Helvetica 53:43-50
Breunig M (1996) Integration of spatial information for Geo-Information Systems (Lecture Notes in Earth Sciences 61). Springer, Berlin/Heidelberg/New York
Breunig M (1999) An approach to the integration of spatial data and systems for a 3D geo-information system. Computers & Geosciences 25:39-48
Brimicombe A (1998) A fuzzy coordinate system for locational uncertainty in space and time. In: Carver S (ed) Innovations in GIS 5. Selected Papers from the Fifth National Conference on GIS Research UK (GISRUK). Taylor & Francis, London/Bristol, pp 141-150
Brockwell PJ, Davis R (1991) Time series. Theory and methods, 2nd ed. Springer, Berlin/Heidelberg/New York
Brodie ML, Mylopoulos J (eds) (1986) On knowledge based information systems. Springer, Berlin/Heidelberg/New York
Brückler M, Dumfarth E (1996) Erstellung von Bodenpreiskarten durch GIS-gestützte räumliche Interpolation. Methode und weiterführende Analysemöglichkeiten. In: Dollinger F, Strobl J (eds) Angewandte Geographische Informationsverarbeitung VIII. Beiträge zum GIS-

Symposium 3.-5. Juli 1996 (Salzburger Geographische Materialien). Institut für Geographie, Universität Salzburg, pp 39-50

Brundsdon C (1995) Analysis of univariate Census Data. In: Openshaw S (eds) Census user's handbook. GeoInformation International, Cambridge, pp 213-237

Brusoni V, Console L, Pernici B, Terenziani P (1997) LaTeR: an efficient, general purpose manager of temporal information. IEEE Expert 12:56-64

BrusoniV, Console L, Terenziani P, Pernici B (1995) Extending temporal relational databases to deal with imprecise and qualitative temporal information. In: Clifford J, Tuzhilin A (eds) Recent advances in temporal databases. Proceedings of the International Workshop on Temporal Databases, Zürich, Switzerland, 17-18 September 1995 (Workshops in Computing). Springer, Berlin/Heidelberg/New York, pp 3-22

Buchmann A (1996) Revisiting common assumptions in real-time databases. Paper presented at the Fifth International Conference on Information and Knowledge Management, Workshop on databases: active and real-time, Rockyville, MD, November 12-16, 1996

Buchroithner M (1999) Möglichkeiten "echter" 3D-Visualisierungen von Geo-Daten. In: Strobl J, Blaschke T (eds) Angewandte Geographische Informationsverarbeitung, Beiträge zum AGIT-Symposium, Salzburg 1999. Wichmann, Heidelberg [http://www.geo.sbg.ac.at /agit/papers99/Buchroithner/buchroithner.htm]

Bund B (1998) Der Wandel der Kulturlandschaft Nordschwarzwald seit der 2. Hälfte des 19. Jahrhunderts. Eine historische Raum-Zeit-Analyse mit Hilfe eines Geographischen Informationssystems (GIS). Mitteilungen der Forstlichen Versuchs- und Forschungsanstalt Baden-Württemberg, Abteilung Betriebswirtschaft. Forstliche Versuchs- und Forschungsanstalt, Freiburg i.Br.

Burns KL (1988) Lithographic topology and structural vector fields applied to sub-surface prediction in geology. Proceedings of GIS/LIS '88, San Antonio, pp 26-34

Burrough P, Van Deursen W, Heuelink G (1988) Linking spatial process models and GIS: a marriage of convenience or a blossoming partnership. Proceedings of GIS/LIS '88, San Antonio, pp 598-608

Burrough PA (1986) Principles of Geographical Information Systems for land resources assessment. Clarendon Press, Oxford

Burrough PA (1992) Are GIS data structures too simple minded? Computers & Geosciences 18:395-400

Burrough PA, Frank AU (1995) Concepts and paradigms in spatial information: are current Geographical Information Systems truly generic? International Journal of Geographical Information Systems 9:101-116

Burrough PA, Frank AU (eds) (1996) Geographic objects with indeterminate boundaries. Taylor & Francis, London

Burrough PA, Frank AU (eds) (1998) European Geographic Information Infrastructures: opportunities and pitfalls (GISData 5). Taylor & Francis, London

Burrough PA, McDonnell RA (1998) Principles of Geographical Information Systems. Oxford University Press, Oxford

Buziek G, Dransch D, Rase WD (eds) (2000) Dynamische Visualisierung von Raum und Zeit: Grundlagen und Anwendungsbeispiele für kartographische Animation. Springer, Berlin/Heidelberg/New York

Calder N (1991) GIS/LIS and the new global geography. Paper presented at GIS/LIS '91, Atlanta

Calkins HW (1984) Space-time data display techniques. In: Marble DF (ed) Proceedings of the 1st International Symposium on Spatial Data Handling. Department of Geography, University of Zurich, Zurich, pp 324-331

Callison HR (1995) A time-sensitive object model for real-time systems. ACM Transactions on software engineering and methodology 4:287-317

Campbell H, Masser I (1995) GIS and organizations. Taylor & Francis, London

Campell JB (1996) Introduction to remote sensing, 2nd ed. Taylor & Francis, London

Canavaggio JF, Dumas M (1997) Manipulation de valeurs temporelles dans un SGBD à objets. Paper presented at 15ième Congrès INFORSID, Toulouse, 1997

Canavaggio JF, Scholl PC, Fauvet MC (1996) Un modèle multigranulaire du temps pour un SGBD temporel. Technical Report RR 962-I, Laboratoire de Logiciels, Systèmes et Réseaux.

Cappelli A, De Castro C, Scalas MR (1995) A modular history-oriented access structure for bitemporal data. Proceedings, 22nd Seminar on Current Trends in Theory and Practice of Informatics SOFSEM '95, pp 369-374

Car A, Frank AU (1994) Hierarchisches räumliches Schließen-Allgemeine Prinzipien. In: Dollinger F, Strobl J (eds) Angewandte Geographische Informationstechnologie VI. Institut für Geographie, Universität Salzburg, pp 151-162

Car A, Frank AU (1995) Formalisierung konzeptioneller Modelle für GIS-Methode und Werkzeug. In: Dollinger F, Strobl J (eds) Beiträge zum GIS-Synopsium 5.-7. Juli 1995. Beiträge zur Geographischen Informationswissenschaft VII. Institut für Geographie, Universität Salzburg, pp 297-305

Carlson E (1987) Three dimensional conceptual modelling of subsurface structures. Proceedings of AutoCarto 8. ASPRS/ACSM, Falls Church, pp 336-345

Carrara A, Guzzetti F (eds) (1995) Geographical Information Systems in assessing natural hazards (Advances in Natural and Technological Hazards Research 5) Kluwer Academic, Dordrecht

Cartwright W, Peterson MP, Gartner G (1999) Multimedia cartography. Springer, Berlin/Heidelberg/New York

Carvalho TP, Edelweiss N (1997a) A visual query system implementing a temporal object-oriented model with roles on a relational database. Proceedings of the 17th International Conference of the Chilean Computer Science Society (SCCC '97), November 12-14, 1997, Valpariso, pp 38-47

Carvalho TP, Edelweiss N (1997b) Um sistema visual de consulta para um modelo de dados temporal orientado a objetos. Anais 24 Seminário Integrado da Sociedade Brasileira de Compucao, pp 337-348

Carver S (ed) (1998) Innovations in GIS 5. Taylor & Francis, London/Bristol

Cassettari S (1993) Introduction to integrated Geo-Information management. Chapman & Hall, London/Glasgow/New York

Castagneri J (1998) Temporal GIS explores new dimensions in time. GIS World 11 [http://www.gisworld.com/gw/1998/0998/998tmp.asp]

Castle GH (ed) (1998) GIS in real estate: integrating, analyzing, and presenting locational information. Appraisal Institute, Chicago

Chakravarthy S (1996) A practical approach to build real-time database applications. Paper presented at the Fifth International Conference on Information and Knowledge Management, Workshop on databases: active and real-time, Rockyville, MD, November 12-16, 1996

Charalampos I, Theodoulidis I, Loucopoulos P (1991) The time dimension in conceptual modelling. Information Systems 16:273-300

Charlton M, Rao L, Carver S (1995) GIS and the census. In: Openshaw S (ed) Census user's handbook. GeoInformation International, Cambridge, pp 133-166

Charlton ME, Openshaw S, Wymer C (1990) Some experiments with an adaptive data structure in the analysis of space-time data. Proceedings of the Fourth International Symposium on Spatial Data Handling, pp 1030-1039

Checkland P (1981) Systems thinking, systems practice. John Wiley & Sons, London

Chen GD, Lee YH, Liu CC (1996) Extending OODB with behavioral temporal management capability. Paper presented at the 20th International Computer Software and Applications Conference COMPSAC '96

Chen HJ, Little TDC (1996) Storage allocation policies for time-dependent multimedia data. IEEE Transactions on Knowledge and Data Engineering 8:855-864

Chen JR, Cheng AMK (1995) Response time analysis of EQL real-time rule-based systems. IEEE Transactions on Knowledge and Data Engineering 7:26-43

Chen PPS (1976) The entity-relationship model: toward a unified view of data. ACM Transactions On Database Systems 1:9-36

Chen YW, Gruenwald L (1996) Effects of deadline propagation on scheduling nested transactions in distributed real-time database systems. Information Systems 21:103-124

Cheng C, Shahar Y, Puerta A, Stites D (1997) Navigation and visualization of abstractions of time-oriented clinical data. Technical Report SMI-97-0688 [http://www. arpuerta.com/pubs/tech97a.pdf]

Cheng TS, Gadia SK (1992) A seamless object-oriented model for spatio-temporal databases. Technical Report TR-92-42. Computer Science Department, Iowa State University, Ames

Cheng TS, Gadia SK (1994) A pattern matching language for spatio-temporal databases. Proceedings of the Conference on Information and Knowledge Management, Gaithersburg, pp 288-295

Cheng TS, Gadia SK, Nair S (1992) Relational and object-oriented parametric databases. Technical Report TR-92-42. Computer Science Department, Iowa State University, Ames

Cheylan JP, Lardon S (1995) Towards a conceptual data model for the analysis of spatio-temporal processes: the example of the search for optimal grazing strategies. In: Frank AU, Kuhn W (eds) Spatial information theory-a theoretical basis for GIS. Proceedings COSIT '95 (LNCS 988). Springer, Berlin/Heidelberg/New York, pp 158-176

Chittaro L, Combi C (1997) Temporal indeterminancy in deductive databases: an approach based on the event calculus. In: Andler S, Hansson J (eds) Active, real-time, and temporal database systems, Second International Workshop, ARTDB-97, Como, Italy, September 8-9, 1997, Proceedings (LNCS 1553). Springer, Berlin/Heidelberg/New York, pp 212-227

Chiu JS, Chen ALP (1996) A note on „incomplete relational database models based on intervals". IEEE Transactions on Knowledge and Data Engineering 8:189-191

Cho G (1998) Geographic Information Systems and the law: mapping the legal frontiers. John Wiley & Sons, Chichester/New York

Chomicki J (1994) Temporal query languages: a survey. Proceedings of the International Conferences on Temporal Logic. Springer, Berlin/Heidelberg/New York, pp 506-534

Chomicki J, Niwinski D (1995) On the feasibility of checking temporal integrity constrains. Journal of Computer and System Science 51:523-535

Chomicki J, Revesz PZ (1997) Constraint-based interoperability of spatiotemporal databases. In: Scholl M, Voisard A (eds) Advances in Spatial Databases, Fifth International Symposium, SSD'97, Berlin, July 15-18, 1997, Proceedings (LNCS 1262). Springer, Berlin/Heidelberg/New York, pp 142-161

Chou YH (1993) Critical issues in the evaluation of spatial autocorrelation. In: Frank AU, Campari I (eds) Spatial information theory. A theoretical basis for GIS. Proceedings COSIT '93 (LNCS 716). Springer, Berlin/Heidelberg/New York, pp 421-433

Chou YH (1995) Spatial pattern and spatial autocorrelation. In: Frank AU, Kuhn W (eds) Spatial information theory-a theoretical basis for GIS. In: Frank AU, Kuhn W (eds) Spatial information theory-a theoretical basis for GIS. Proceedings COSIT '95 (LNCS 988). Springer, Berlin/Heidelberg/New York, pp 365-376

Chrisman N (1983) The role of quality information in the long-term functioning of a Geographic Information System. Proceedings of AutoCarto 6, pp 303-321

Chrisman N (1984) The role of quality information in the long-term functioning of a Geographic Information System. Cartographica 21:79-87

Chrisman N (1997) Exploring Geographic Information Systems. John Wiley & Sons, New York

Chrisman NR (1993) Beyond spatio-temporal data models: a model of GIS as a technology embedded in historical context. Proceedings AutoCarto 11, pp 23-32

Chrisman NR (1998) Beyond the snapshot: changing the approach to change, error, and process. In: Egenhofer MJ, Golledge RG (eds) Spatial and temporal reasoning in Geographic Informations Systems. Oxford University Press, New York/Oxford, pp 85-93

Claramunt C, Bai L (1999) A multi-scale approach to the propagation of temporal constraints in GIS. Journal of Geographic Information and Decision Analysis 3:9-20

Claramunt C, Bargiela A, Peytchev E (1999) A real-time GIS for the analysis of a traffic system. Proceedings of the 5th IEEE African '99 Conference, Cape Town, pp 15-20

Claramunt C, De Sède MH, Prelaz-Drou R, Vidale L (1994) Sémantique et logique spatio-temporelles. Revue Internationale de Geomatique 4:165-180

Claramunt C, Jiang B (2000) A representation of relationships in temporal spaces. In: Atkinson P, Martin D (eds) Innovations in GIS VII: GeoComputation. Taylor & Francis, London, pp 41-53

Claramunt C, Parent C, Thériault M (1997) Design patterns for spatio-temporal processes. In: Spaccapietra S, Maryanski F (eds) Searching for semantics: data mining, reverse engineering. Chapman & Hall, pp 415-428

Claramunt C, Thériault M (1995) Managing time in GIS. An event-oriented approach. In: Clifford J, Tuzhilin A (eds) Recent advances in temporal databases. Proceedings of the International Workshop on Temporal Databases, Zurich, Switzerland, 17-19 September 1995. Springer, Berlin/Heidelberg/New York, pp 23-42

Claramunt C, Thériault M (1996) Toward semantics for modelling spatio-temporal processes within GIS. In: Kraak MJ, Molenaar M (eds) Advances in GIS II. Taylor & Francis, Delft, pp 47-64

Claramunt C, Thériault M, Parent C (1997) A qualitative representation of evolving spatial entities in two dimensional topological spaces. In: Carver S (ed) Innovations in GIS V. Taylor & Francis, London, pp 129-139

Clark M et al. (1992) Waste location: spatial aspects of waste management, hazards and disposal (The Natural Environment: Problems and Management Series). Routledge, London

Clarke KC (1999) Getting started with Geographic Information Systems, 2nd ed. Prentice Hall, Upper Saddle River

Clementini E, Di Felice P, Van Oosterom P (1993) A small set of formal topological relationships suitable for end-user interaction. In: Abel D, Ooi BC (eds) Advances in spatial databases, Third International Symposium SSD '93 (LNCS 692). Springer, Berlin/ Heidelberg/New York, pp 277-295

Clifford J, Croker A, Grandi F, Tuzhilin A (1995) On temporal grouping. In: Clifford J, Tuzhilin A (eds) Recent advances in temporal databases. Proceedings of the International Workshop on Temporal Databases, Zürich, Switzerland, 17-18 September 1995 (Workshops in Computing). Springer, Berlin/Heidelberg/New York, pp 194-213

Clifford J, Dyreson C, Isakowitz T, Jensen CS, Snodgrass RT (1997) On the semantics of „now" in databases. ACM Transactions on Database Systems 22:171-214

Clifford J, Jensen CS, Snodgrass RT, Böhlen MH, Dewan H, Schmidt D (1995) Panel: the state-of-the-art in temporal data management: perspectives from the research and financial applications communities. In: Clifford J, Tuzhilin A (eds) Recent advances in temporal databases. Proceedings of the International Workshop on Temporal Databases, Zürich, Switzerland, 17-18 September 1995 (Workshops in Computing). Springer, Berlin/Heidelberg/New York, pp 336-357

Clifford J, Tuzhilin A (1995) Recent advances in temporal databases. In: Clifford J, Tuzhilin A (eds) Recent advances in temporal databases. Proceedings of the International Workshop on Temporal Databases, Zürich, Switzerland, 17-18 September 1995 (Workshops in Computing). Springer, Berlin/Heidelberg/New York

Codd E (1970) A relational model for large shared data banks. ACM Communications 13:377-387

Codd EF (1986) Cellular automata. Academic Press, New York

Cohen M, Alexandre F, Mathieu N (1998) Modelling interactions between biophysical and social systems: The example of dynamics of vegetation in Causse Mejan (France). Cybergeo 39 [http://193.55.107.3/revgeo/rostok/textes/cohen.htm]

Cole J, King CAM (1969) Quantitative geography: techniques and theories in geography. John Wiley & Sons, New York

Cole S (1998) Of maps and macros: object-oriented spreadsheet GIS. Environment and Planning B 25:227-243

Combi C, Cucchi G (1997) GCH-OSQL: a temporally-oriented object-oriented query language based on a three-valued logic. Proceedings of the Fourth International Workshop on Temporal Representation and Reasoning TIME '97, pp 119-126

Combi C, Cucchi G, Pinciroli F (1997) Applying object-oriented technologies in modeling and querying temporally-oriented clinical databases dealing with temporal granularity and indeterminancy. IEEE Transactions on Information Technology in Biomedicine 1:100-127

Combi C, Pinciroli F, Pozzi G (1995) Managing different time granularities of clinical information by an interval-based temporal data model. Methods of Information in Medicine 34:458-474

Combi C, Pinciroli F, Pozzi G (1996) Managing time granularity of narrative clinical information: the temporal data model TIME-NESIS. In: Chittaro L et al. (eds) Proceedings of the Third International Workshop on Temporal Representation and Reasoning TIME '96. IEEE, Washington, pp 88-93

Combi C, Shahar Y (1997) Temporal reasoning and temporal data maintenance in medicine: issues and challenges. Computers in Biology and Medicine 27: 353-368

Conrad S (1996) Report on the international workshop on temporal reasoning in deductive and object-oriented databases. The Knowledge Engineering Review 11:191-192

Coomber M (1995) Dealing with census geography: principles, practices and possibilities. In: Openshaw S (ed) Census user's handbook. GeoInformation International, Cambridge, pp 111-132

Couclelis H (1985) Cellular worlds: a framework for modelling micro-macro dynamics. Environment and Planning A 17:585-596

Couclelis H (1986) A theoretical framework for alternative models of spatial decisions and behaviour. Annals of the Association of American Geographers 76:95-113

Couclelis H (1993) Location, place, region and space. In: Abler RF, Marcus MG, Olson JM (eds) Geography's inner worlds: pervasive themes in contemporary American geography. Rutgers University Press, New Brunswick, pp 215-233

Couclelis H (1996) Geographic illusion systems: towards a (very partial) research agenda for GIS in the information age. Position paper, Specialist meeting on GIS and society [http://www.geo.wvu.edu/i19/papers/couclelis.html]

Couclelis H (1999) Space, time, geography. In: Longley PA et al. (eds) Geographical Information Systems, principles and technical issues, 2nd ed. John Wiley & Sons, New York, pp 29-38

Cowen DJ, Shirley WL (1992) Integrated planning information systems. In: Maguire DJ, Goodchild MF, Rhind DW (eds) Geographical Information Systems, Applications. Longman, Harlow, pp 297-310

Cressie NAC (1991) Statistics for spatial data. John Wiley & Sons, New York

Cromley RG (1983) Automated geography: some problems and pitfalls. The Professional Geographer 35:340-341

Crosbie P (1996) Object-oriented design of GIS. Computional modeling systems to support the development of scientific models. In: Goodchild MF et al. (eds) GIS and environmental modeling: progress and research issues. GIS World Books, Fort Collins, pp 383-386

Cross AE (1991) Who said GIS cannot fail? Mapping Awareness 5:10

Cu WW, Ieong IT, Taira RK (1994) A semantic modeling approach for image retrieval by content. VLDB Journal 3:445-478

Cukierman D, Delgrande J (1996) Characterizing temporal repetition. In: Chittaro L et al. (eds) Proceedings of the Third International Workshop on Temporal Representation and Reasoning TIME '96. IEEE, Washington, pp 80-87

Curry MR (1991) The architectonic impulse and the recoceptualization of the concrete in contemporary geography. In: Barnes T, Duncan J (eds) Writing worlds: discourse, text and metaphor in the representation of landscape. Routledge, London, pp 97-117

Curry MR (1994) Image, practice and the hidden impacts of Geographic Information Systems. Progress in Human Geography 18:441-459

Curry MR (1995) On the inevitability of ethical inconsistency in Geographical Information Systems. In: Pickles J (ed) Ground truth: the social implications of Geographic Information Systems. Guilford Press, New York, pp 68-87

Curry MR (1998) Digital places: living with Geographic Information Technologies. Routledge, London

Dababrata D, Barron TM, Storey VC (1996) A complete temporal relational algebra. The VLDB Journal 5:167-180

Daniel L, Whitener A, Loree P (1999) Inside MapInfo professional, 2nd ed. OnWord Press, Albany

Dao S (1997) Modeling and querying moving objects. Proceedings of the 13th International Conference on Data Engineering ICDE 13, pp 422-432

Daroczi Z (1998) Geokodierung von Kundenadressen–der Schlüssel zum erfolgreichen Geomarketing, dargestellt am Beispiel Österreich. Geographische Informationssysteme 11:20-23

Darwen H (1995) Normalising temporal tables. Technical Report, ISO/IEC JTC 1/SC 21/WG 3 DBL LHR-33r2

Darwen H (1996) Temporal SQL: Support for data type period. Technical Report, ISO/IEC JTC 1/SC 21/WG 3 DBL YOW-23

Date CJ (1995) An introduction to database systems, 6th ed. Addison-Wesley, Reading

Datta A, Mukherjee S, Konana P, Viguier IR, Bajaj A (1996) Multiclass transaction scheduling and overload management in firm real-time database systems. Information Systems 21:29-54

David B, Raynal L, Schroter G (1993) GeO2: Why objects in a geographical DBMS? In: Abel D, Ooi BC (eds) Advances in spatial databases, Third International Symposium SSD 93 (LNCS 692). Springer, Berlin/Heidelberg/New York, pp 264-276

Davies C, Lazell B, Hughes M, Cooper L (1995) Time is just another attribute-or at least, just another dimension. In: Clifford J, Tuzhilin A (eds) Recent advances in temporal databases. Proceedings of the International Workshop on Temporal Databases, Zürich, Switzerland, 17-18 September 1995 (Workshops in Computing). Springer, Berlin/Heidelberg/New York, pp 175-193

Davis BE (1996) Geographic Information Systems: a visual approach. OnWord Press, Santa Fe

Davis BE, Williams R (1989) The five dimensions of GIS. Proceedings GIS/LIS'89, Washington, DC, pp 50-58

Davis DE (1999) GIS for everyone: exploring your neighborhood and your world with a Geographic Information System. ESRI Press, Redlands

De Boer F (1990) Spreadsheets and GIS in integral modeling. In: Pillmann W, Jaeschke A (eds) Informatik für den Umweltschutz. 5. Symposium Wien, 19.-21. September 1990 (Informatik Fachbericht). Springer, Berlin/Heidelberg/New York, pp 549-556

De Castro C (1996a) Levels of concurrency control in temporal relational databases supporting schema visioning. Technical Report CSITE-n.ijk-96.

De Castro C (1996b) Temporal conversion functions for multitemporal relational databases. Proceedings of the Third Seminar on Current Trends in Theory and Practice of Informatics (SOFSEM '96), pp 343-350

De Castro C, Grandi F, Scalas MR (1995) On schema versioning in temporal databases. In: Clifford J, Tuzhilin A (eds) Recent advances in temporal databases. Proceedings of the International Workshop on Temporal Databases, Zürich, Switzerland, 17-18 September 1995 (Workshops in Computing). Springer, Berlin/Heidelberg/New York, pp 272-291

De Castro C, Grandi F, Scalas MR (1996) Correttezza e ottimizzazione delle interrogazioni nelle basi di dati relazionali multitemporali. Proceedings of Qu Conferenza Nationale su Sistemi Evoluti per Basi di Dati SEBD '96, pp 53-71

De Castro C, Grandi F, Scalas MR (1997) Schema versioning for multitemporal relational databases. Information Systems 22:249-290

De Floriani L, Marzano P (1993) Spatial queries and data models. In: Frank AU, Campari I (eds) Spatial information theory. A theoretical basis for GIS. Proceedings COSIT '93 (LNCS 716). Springer, Berlin/Heidelberg/New York, pp 113-138

De Floriani L, Marzano P (1995) Spatial queries and data models. In: Frank AU, Kuhn W (eds) Spatial information theory-a theoretical basis for GIS. Proceedings COSIT '95 (LNCS 988). Springer, Berlin/Heidelberg/New York, pp 113-137

De Hoop S, Van Oosterom P, Molenaar M (1993) Topological querying of multiple map layers. In: Frank AU, Campari I (eds) Spatial information theory. A theoretical basis for GIS. Proceedings COSIT '93 (LNCS 716). Springer, Berlin/Heidelberg/New York, pp 139-157

De Jong T, Ottens H (1997) GIS functionality for multi-level research. In: Westert GP, Verhöeff RN (eds) Places and people: multilevel modelling in geographical research (Nederlandse Geografische Studies). KNAG, Utrecht, pp 44-54

De Jong T, Ritsema van Eck J (1997) Netwerkanalyse, bereikbaarheidsanalyse en GIS. In: Hendriks P, Ottens H (eds) Geografische Informatie Systemen in ruimtelijk onderzoek. Van Gorcum, Assen, pp 145-164

De Lange N (1997) Grundlagen von Geo-Informations-Systemen: theoretische Grundlagen und Einführung in ARC/INFO anhand einer Projektaufgabe. Universität Osnabrück, Fachgebiet Geographie, Osnabrück

De Lepper MJC et al. (1995) The added value of Geographical Information Systems in public and environmental health (Geojournal Library 24). Kluwer, Dordrecht

Demel W, Franz H, Kias U (1996) Natur- und Biotopschutz. Lektion 8: Landnutzung und Landschaftswandel im Biosphärenreservat Berchtesgaden. In: Bumann E, Bachhuber R, Schaller J (eds) ArcView. GIS-Arbeitsbuch. Wichmann, Heidelberg, pp 147-161

DeMers MN (1997) Fundamentals of Geographic Information Systems. John Wiley & Sons, New York

Deutsches Nationalkomitee für das Unesco Programm „Der Mensch und die Biosphäre" (MAB) (ed) (1990) Geographische Informationssysteme und ihre Anwendung in MAB-Projekten, Ökosystemforschung und Umweltbeobachtung (MAB-Mitteilungen 34). MAB, Bonn

Di Biase D, MacEachren AM, Krieger JB, Reeves C (1992) Animation and the role of map design in scientific visualization. Cartography and Geographic Information Systems 19:201-214

Dibble C (1998) Representing individuals and societies in GIS [http://www.geo.wvu.edu/i19/papers/dibble.html]

Dickmann F (1999) Computerkartographie und GIS. Westermann, Braunschweig

Didinger O (1996) Bereitstellung von Basisinformationen: automatisiertes Liegenschaftsbuch ALB, automatisierte Liegenschaftskarte ALK, Amtliches topographisch-kartographisches Informationssystem ATKIS. Landesvermessungsamt Rheinland-Pfalz (ed) Das Geoinformationssystem ATKIS und seine Nutzung in Wirtschaft und Verwaltung. LVA Rheinland-Pfalz, Koblenz, pp 19-30

Diggle PJ (1983) Statistical analysis of spatial point pattern. Academic Press, London

Dignum F, Weigand H, Verharen E (1996) Meeting the deadline: on the formal specification of temporal denotic constraints. In: Ras Z, Michalewicz M (eds) Proceedings of the 9th International Symposium on Foundations of Intelligent Systems (ISMIS '96), pp 243-252

Ding C (1998) The GIS-based human-interactive TAZ design algorithm: examining the impacts of data aggregation on transportion-planning analysis. Environment and Planning B 25:601-616

Dionisio JDN, Cardenas, AF (1996) Mquery: a visual query language for multimedia, timeline and simulation data. Journal of Visual Languages and Computing 7:377-401

DiPippo LC, Wolfe VF (1997) Object-based semantic real-time concurrency control with bounded imprecision. IEEE Transactions on Knowledge and Data Engineering 9:135-147

Dobson J (1983) Automated Geography. The Professional Geographer 35:135-143

Dodge M (1996) The visualisation and analysis of fire incidents using GIS. Proceedings of the Joint European Conference and Exhibition on Geographical Information, March 27-29th 1996, Barcelona, pp 592-601

Dogdu E, Özsoyoglu G (1996) Adaptive real-time transactions and risk-based load control. Paper presented at the Fifth International Conference on Information and Knowledge Management, Workshop on databases: active and real-time, Rockyville, MD, November 12-16, 1996

Dollinger F (1989) Wie kam die Geographie zum GIS? Über die Entwicklung des GIS-Konzeptes in der Landschaftsforschung. In: Dollinger F, Strobl J (eds) Beiträge zum GIS-Synopsium 5.-7. Juli 1989. Angewandte Geographische Informationstechnologien I (Salzburger Geographische Materialien). Institut für Geographie, Universität Salzburg, pp 11-26

Dollinger F (1992) Geoinformatik. Einige Randbemerkungen zur Entwicklung einer jungen Wissenschaft. In: Dollinger F, Strobl J (eds) Beiträge zum GIS-Synopsium 8.-10. Juli 1992. Angewandte Geographische Informationstechnologien IV (Salzburger Geographische Materialien). Institut für Geographie, Universität Salzburg, pp 7-10

Döllner J, Kersting O, Hinrichs K, Baumann K (1999) Konzepte und 3D-Visualisierung interaktiver, perspektivischer Karten. In: Strobl J, Blaschke T (eds) Angewandte Geographische Informationsverarbeitung, Beiträge zum AGIT-Symposium, Salzburg 1999. H. Wichmann, Heidelberg [http://www.geo.sbg.ac.at/agit/papers99/Doellner/doellner.htm]

Domosh M (1992) Toward a feminist historiography of geography. Transactions, Institute of British Geographers 16:95-104

Dorling D (1992) Stretching space and splicing time: from cartographic animation to interactive visualization. Cartography and Geographic Information Systems 19:215-227

Dorling D (1995) Visualizing the 1991 census. In: Openshaw S (ed) Census user's handbook. GeoInformation International, Cambridge, pp 167-211

Dorling D, Openshaw S (1992) Using computer animation to visualise space-time patterns. Environment and Planning B 19:639-650

Doucet A, Fauvet MC, Gancarski S, Jomier G, Monties S (1997) Using database versions to implement temporal integrity constrains. Proceedings of the Second International Workshop on Constraint Database Systems CDB '97. (LNCS 1191), pp 219-233

Douglas WJ (1995) Environmental GIS: applications to industrial facilities. Lewis Publishers, Boca Raton/London

Dransch D (1997) Computer-Animation in der Kartographie. Theorie und Praxis. Springer, Berlin/Heidelberg/New York

Dumfarth E (1996) Interpolation von Oberflächen mit CRS (completely regularized spline with smoothing and tension)-Evaluierung von Methode und Ergebnisse am Beispiel „Bodenpreisoberflächen für die Stadt Salzburg“. In: Dollinger F, Strobl J (eds) Angewandte Geographische Informationsverarbeitung VIII. Beiträge zum GIS-Symposium 3.-5. Juli 1996 (Salzburger Geographische Materialien). Institut für Geographie, Universität Salzburg, pp 27-38

Dutta K, Perkins WA (1990) Inferring attribute values from spatial and temporal representations. In: Trivedi MM (ed) Applications of artificial intelligence VIII. SPIE, Bellingham, pp 558-571

Dutta S (1991) Topological constraints: a representational framework for approximate spatial and temporal reasoning advances in spatial databases. 2nd Symposium SSD '91 (LNCS). Springer, Berlin/Heidelberg/New York, pp 161-180

Dutton GH (1999) A hierarchical coordinate system for geoprocessing and cartography (Lecture Notes in Earth Sciences 79). Springer, Berlin/Heidelberg/New York

Dyreson CE, Snodgrass RT (1992) Time-stamp semantics and representations [ftp://ftp.cs.arizona.edu/reports/1992/TR92-16.ps]

Dyreson CE, Snodgrass RT (1998) Supporting valid-time indeterminacy. ACM Transactions on database systems 23:1-57

Eagles M (ed) (1995) Spatial and contextual models in political research. Taylor & Francis, London

Easterfield M, Newell R, Thériault D (1990) Management of time and space. Paper presented at the AGI '90 Conference

Easterfield M, Newell RG, Thériault D (1991) Modelling spatial and temporal information. Proceedings of EGIS '91, Brussels, pp 294-304

Eastman, JR (1985) Cognitive models and cartographic design. Cartographic Journal 22:95-101

Ebbinghaus J (1994) Einsatz objektorientierter Datenbanktechnologie im geographischen Informationssystem GODOT. In: Kremers H (ed) Umweltdatenbanken. Metropolis-Verlag, Marburg, pp 37-58

Ebdon D (1992) SPANS-A quatree-based GIS. Computers & Geosciences 18:471-475

Ebeling D (ed) (1999) Historisch-thematische Kartographie: Konzepte–Methoden–Anwendungen. Verlag für Regionalgeschichte, Bielefeld

Ebeling D, Freimuth P, Nagel JG, Schmidt M, Stein T (1999) ARASS–Ein Datenmanagementsystem als Grundlage eines offenen Geoinformationssystems mit dem Schwerpunkt auf den hisorischen Wissenschaften. In: Ebeling D (ed) Historisch-thematische Kartographie: Konzepte – Methoden – Anwendungen. Verlag für Regionalgeschichte, Bielefeld, pp 181-196

Edelweiss N, Oliveira JPM (1996) Integrating DFD and ER models using a temporal object-oriented model with roles. Proceedings of Engineering Systems Design and Analysis Conference, pp 173-181

Edelweiss N, Oliveira JPM (1997) Multidimensional enterprise modelling using a temporal object-oriented model with roles. Proceedings of the First East-European Symposium on Advances in Databases and Information Systems, pp 394-398

Edsall R, MacEachren AM, Kraak MJ (1998) The use of temporal legends by users of interactive visualization systems. Paper presented at the Annual Meeting of the Association of American Geographers, Fort Worth, March 2-29, 1998

Edsall R, Peuquet D (1996) Graphical query techniques for temporal GIS. Paper submitted for the 13th AutoCarto Conference, Seattle, 1997 [http://www.geog.psu.edu/~edsall/research/autocarto.html]

Edsall RM, Harrower M, Mennis J (2000) Visualizing properties of spatial and temporal periodicity in geographic data. Computers and Geosciences 26:109-118

Edsall RM, Peuquet DJ (1997) A graphical user interface for the integration of time into GIS. Proceedings of the 1997 American Congress of Surveying and Mapping Annual Convention and Exhibition, Seattle, pp 182-189 [http://www.geo.psu.edu/~edsall/research/ACarticle.html]

Egenhofer M, Al-Taha K (1992) Reasoning about gradual changes of topological relationships In: Frank AU, Campari I, Formentini U (eds) Theories and methods of spatio-temporal reasoning in geographic space (LNCS), pp 169-219

Egenhofer MJ, Golledge RG (eds) (1998) Spatial and temporal reasoning in Geographic Information Systems (spatial information systems). Oxford University Press, Oxford/New York

Egenhofer MJ, Herring JR (1991) High-level spatial data structures for GIS. In: Maguire DJ, Goodchild MF, Rhind DW (eds) Geographical Information Systems. John Wiley & Sons, New York, pp 227-237

Egenhofer MJ, Sharma J (1993) Topological relations between regions in R2 and R3. In: Abel D, Ooi BC (eds) Advances in spatial databases, Third International Symposium SSD '93, Proceedings (LNCS 692). Singapore, pp 316-336

Eini I, Goebel V, Skjellaug B (1997) A temporal data model for multimedia database systems Technical Report, Research Report 252. University of Oslo, Department of Informatics, Oslo

Ellis S (1994) GIS 2000BC-AD2000. AGI '94 Conference Papers Association for Geographic Information. AGI, London, pp 12.2.1-12.2.5

Elmarsi R, Wuu GTJ, Kim YJ (1990) The time index: an access structure for temporal data. Proceedings of the 16th VLDB Conference. VLDB, Brisbane, pp 1-11

Environmental Systems Research Institute (ESRI) (1995) Managing data history Arc/Info version 7.03 on-line help file. ESRI Press, Redlands

Environmental Systems Research Institute (ESRI) (1997) GIS data storage trends [http://www.esri.com/library/whitepapaers/pdfs/datastor_util.pdf]

Environmental Systems Research Institute (ESRI) (1999) Getting to know ArcView GIS, 3rd ed. ESRI Press, Redlands

Ernst E, Herderich C, Weisemann U et al. (1995) Metainformation zu Messdaten aus der Sicht von Simulationsmodellen. Zwischenergebnisse aus dem UFIS Projekt zur Entwicklung einer Metadatenbank über Daten. In: Kremers H, Pillmann W (eds) Raum und Zeit in Umweltinformationssystemen. 9th International Symposium on Computer Science for Environmental Protection CSEP '95 (Umwelt-Informatik aktuell). Metropolis-Verlag, Marburg, pp 408-413

Erwig M et al. (1997) Spatio-temporal data types: an approach to modelling and querying moving objets in databases. CHOROCHRONOS-a network for spatiotemporal database systems. Technical Report CH-97-8 [http://www.dbnet.ntua.gr/~choros/TRs/1997/8/report.ps.gz]

Erwig M, Schneider M (1997) Partition and conquer. CHOROCHRONOS-A network for spatiotemporal database systems. Technical Report CH-97-7 [http://www.dbnet.ntua.gr/~choros/TRs/1997/7/report.ps.gz]

Etches A, Claramun C, Bargiela A, Kosonen I (1999) An interoperable TGIS model for traffic systems. In: Gittings B (ed) Innovations in GIS 6, Integrating information infrastructures with GI technology. Taylor & Francis, London, pp 217-228

Fauerbach E, Edsall R, Barnes D, MacEachren A (1996) Visualization of uncertainty in metrological forecast models. Proceedings of the 7th International Spatial Data Handling Conference, Delft, August 12-16, pp 17-28 [http://www.geovista.psu.edu/ica/icavis/febm.html]

Fauvet MC, Canavaggio JF, Scholl PC (1996) Expressions de requêtes temporelles dans un SGBD à objects. Actes des 12ième journées bases de données avancées, Cassis, pp 225-250

Fauvet MC, Canavaggio JF, Scholl PC (1997a) A proposal for histories in object DBMS. Actes des 13ième journèes de bases de données avancées, Grenoble, France. Springer, Berlin/Heidelberg/New York, pp 112-121

Fauvet MC, Canavaggio JF, Scholl PC (1997b) Modeling histories in object DBMS. In: Hameurlain A, Tjoa AM (eds) Database and Expert Systems Applications, 8th International Conference, DEXA '97, Toulouse, September 1-5, 1997 (LNCS 1308), pp 112-121

Fauvet MC, Chardonnel S, Dumas M, Scholl PC, Dumolard P (1998) Analyse de données géographiques: application des bases de données temporelles. Revue Internationale de Géomatique 8

Fedra K (1990) A computer-based approach to environmental impact assessment. In: Colombo A, Premazzi G (eds) Proceedings of the Workshop on Indicators and Indices for Environmental Impact Assessment and Risk Analysis. Office for Official Publications of the European Communities, Luxembourg, pp 11-40

Fedra K (1993) GIS and environmental modeling. In: Goodchild MF, Parks BO, Steyaert LT (eds) Environmental modeling with GIS. Oxford University Press, New York, pp 35-50

Fernandes S, Schiel U, Catarci T (1997) Using formal operators for visual queries in historical databases. Proceedings of the XII Simposio Brasileiro de Banco de Dados, pp 375-387

Fernandes Silva S, Schiel U, Catarci T (1997) Visual query operators for temporal databases. Proceedings of the Fourth International Workshop on Temporal Representation and Reasoning, TIME '97, May 10-11, 1997, Daytona Beach, pp 46-53

Finger M, McBrien P (1997) Concurrency control for perceivedly instantaneous transactions in valid-time databases. Proceedings of the Fourth International Workshop on Temporal Representation and Reasoning, TIME '97, May 10-11, 1997, Daytona Beach, pp 112-118

Fischer ME, Scholten HJ, Unwin A (1996) Geographic Information Systems, spatial data analysis and spatial modelling: an introduction. In: Fischer ME, Scholten HJ, Unwin D (eds) Spatial analytical perspectives on GIS (GISData 4). Taylor & Francis, London/Bristol, pp 3-19

Fischer MM, Nijkamp P (eds) (1993) Geographic Information Systems, spatial modelling, and policy evaluation. Springer, Berlin/Heidelberg/New York

Fischer MM, Unwin D (ed) (1996) Spatial analytical perspectives on GIS. Taylor & Francis, London

Fisher P (1989) Expert system applications in human geography. Area 21:279-287

Flewelling D, Frank A, Egenhofer M (1992) Constructing geological cross sections with a chronology of geological events. Proceedings of the Fifth International Symposium on Spatial Data Handling, pp 544-553

Fliedner D (1987) Raum, Zeit und Umwelt-eine theoretische Betrachtung aus anthropogeographischer Sicht. Geographische Zeitschrift 75:72-85

Floris FJT, Ritsema IL (1994) Spatio-temporal (4D, Dynamic) subsurface models. In: Hack R, Sides E (eds) 3-dimensional GIS: recent developments. ITC Journal 1:66-67

Flowerdew R, Green M (1995) Areal interpolation and types of data. In: Fotheringham S, Rogerson P (eds) Spatial analysis and GIS. Taylor & Francis, London/Bristol, pp 121-145

Flowerdew R, Green M, Kehris E (1991) Using areal interpolation methods in Geographic Information Systems. Papers in Regional Science 70:303-315

Forer P (1998) Geometric approaches to the nexus of time, space, and microprocess: implementing a practical model for mundane socio-spatial systems. In: Egenhofer MJ, Golledge RG (eds) Spatial and temporal reasoning in Geographic Informations Systems. Oxford University Press, New York/Oxford, pp 171-190

Foresman TW (ed) (1998) The history of Geographic Information Systems: perspectives from the pioneers (Prentice Hall Series in Geographic Information Science). Prentice Hall, Upper Saddle River

Förstner W, Plümer, L (eds) (1997) Semantic modeling for the acquisition of topographic information from images and maps. Birkhäuser, Basel/Boston

Fotheringham S, Rogerson P (eds) (1994) Spatial analysis and GIS (technical issues in Geographic Information Systems). Taylor & Francis, London

Fotheringham S, Wegner M (2000) Spatial models and GIS: new potentials and new models GISData 7. Taylor & Francis, London

Fotheringham SA (1998) Trends in quantitative methods II: stressing the computational. Progress in Human Geography 22:283-292

Frank A (1988) Requirements for a database management system for a GIS. Photometric Engineering and Remote Sensing 54:1557-1564

Frank AU (1992) Spatial concepts, geometric data models, and geometric data structures. Computers & Geosciences 18:409-412

Frank AU (1996) An object-oriented, formal approach to the design of castradal systems. In: Kraak MJ, Molenaar M (eds) Proceedings of 7th International Symposium on Spatial Data Handling, SDH '96, Advances in GIS Research II. Taylor & Francis, London, pp 5A.19-5A.35

Frank AU (1998) Different types of "Times" in GIS. In: Egenhofer MJ, Golledge RG (eds) Spatial and temporal reasoning in Geographic Information Systems. Oxford University Press, New York/Oxford, pp 40-62

Frank AU et al. (2000) Life and motion of socio-economic units (GISData 8). Taylor & Francis, New York

Frank AU, Campari I, Formentini U (eds) (1992) GIS–from space to territory: theories and methods of spatio-temporal reasoning in geographic space. International Conference, 21-23 September Pisa, Italy, Proceedings. Springer, Berlin/Heidelberg/New York

Frank AU, Egenhofer MJ (1992) Computer cartography for GIS: an object-oriented view on the display transformation. Computers & Geosciences 18:975-987

Frank AU, Kuhn W (1986) Cell graphs: a provable correct method for the storage of geometry. In: Marble D (ed) Proceedings of the Second International Symposium on Spatial Data Handling, Williamsville, pp 411-436

Frank AU, Mark DM (1991) Language issues for GIS. In: Maguire DJ et al. (eds) Geographical Information Systems, principles and technical issues. John Wiley & Sons, New York, pp 147-163

Freeston M (1987) The BANG file: a new kind of grid file. In: Dayal U, Traiger I (eds) Proceedings of the SIGMOD '87. ACM, San Fransisco, pp 260-269

Freeston M (1989) A well-behaved file structure for the storage of spatial objects. In: Buchmann AP, Günther O, Smith TR, Wang YF (eds) Design and implementation of large spatial databases, First Symposium SSD'89, Santa Barbara, July 17/18, 1989. Proceedings (LNCS 409). Springer, Berlin/Heidelberg/New York, pp 287-300

Freska C (1992) Temporal reasoning based on semi-intervals. Artificial Intelligence 54:199-227

Frevel H (1996) Konzeption, Aufbau und Datenangebot des Geoinformationssystems ATKIS. Landesvermessungsamt Rheinland-Pfalz (ed) Das Geoinformationssystem ATKIS und seine Nutzung in Wirtschaft und Verwaltung. LVA Rheinland-Pfalz, Koblenz, pp 31-46

Fritsch D, Anders KH (1996) Objektorientierte Konzepte in Geo-Informationssystemen. GIS 2:2-14

Frost JD, Chameau JLA (eds) (1993) Geographic Information Systems and their application in geotechnical earthquake engineering: proceedings of a workshop. Georgia. American Society of Civil Engineers, New York

Gabbay D, McBrien P (1991) Temporal logic and historical databases. In: Lohmann GM, Sernadas A, Champs R (eds) 17th International Conference on Very Large Data Bases, September 3-6, 1991, Barcelona, Proceedings. Morgan Kaufmann, pp 423-430

Gadia SK (1988) A homogenous relational model and query languages for temporal databases. ACM Transactions on Database Systems 13:418-448

Gadia SK (1993) Seamless integration of spatial, temporal, belief and ordinary data. SIGMOD Record 22:15-20

Gadia SK, Chopra V (1992) A relational model and SQL-like language for seamless query of temporal data. Technical Report, TR-92-05. Computer Science Department, Iowa State University, Ames

Gadia SK, Chopra V (1993) A relational model and SQL-like query language for seamless query of temporal data. In: Adam NR, Bhargava BK (eds) Advanced database systems (LNCS 759). Springer, Berlin/Heidelberg/New York pp 213-225

Gadia SK, Chopra V, Nair S, Tim US (1992) An SQL-like seamless query language and optimization of spatio-temporal data. Technical Report, TR-92-40. Department of Computer Science, Iowa State University, Ames

Gadia SK, Chopra V, Tim US. (1992) Seamless SQL-like query of spatiotemporal data and a case study in agricultural environment management TR-92-21. Computer Science Department, Iowa State University, Ames

Gadia SK, Nair S (1993) Temporal databases: a prelude to parametric data. In: Tansel A, Clifford J, Gadia SK, Jajodia S, Segev A, Snodgrass RT (eds) Temporal databases: theory, design, and implementation. Benjamin Cummings, Redwood City, pp 28-66

Gadia SK, Vaishnav JH (1985) A query language for a homogenous temporal database. Proceedings of the ACM Symposium on Principles of Database Systems, pp 51-56

Gadia SK, Yeung CS (1988) A generalized model for a relational temporal database. Proceedings of ACM SIGMOD International Conference on Management of Data, pp 251-259

Gaffney V, Stancic Z (1991) GIS approaches to regional analysis. Znanstveni Institut, Filozofske Fakultete, Ljubljana

Gagne JR, Plaice J (1996) A non-standard temporal deductive database system. Journal of Symbolic Computation 22:649-664

Gaile S, Olsson G (eds) (1979) Philosophy in geography. Reidel, Dordrecht

Gal A, Dori D (1996) Combining simultaneous values and temporal data dependencies. In: Chittaro L et al. (eds) Proceedings of the Third International Workshop on Temporal Representation and Reasoning TIME '96. IEEE, Washington, pp 112-119

Gal A, Etzion O (1997) Parallel execution model for updating temporal databases. International Journal of Computer Systems Science and Engineering 12:317-327

Gal A, Etzion O, Segev A (1995) A language for the support of constrains in temporal active databases. Proceedings Workshop on Constrains, Databases and Logic Programming, Portland, pp 42-58

Gal A, Etzion O, Segev A (1996) TALE-a temporal active language and execution model. In: Constantopoulos P, Mylopoulos J, Vassiliou Y (eds) Advanced Information Systems Engineering. Springer, Berlin/Heidelberg/New York, pp 60-81

Gallop J (1994) State of the art in visualization software. In: Hearnshaw HM, Unwin DJ (eds) Visualization in GIS. John Wiley & Sons, Chichester, pp 40-48

Galton A (1995) Towards a qualitative theory of movement. In: Frank AU, Kuhn W (eds) Spatial information theory-a theoretical basis for GIS. Proceedings COSIT '95 (LNCS 988). Springer, Berlin/Heidelberg/New York, pp 377-396

Galton A (1997a) Continuous in spatial regions. In: Hirtle SC, Frank AU (eds) Spatial information theory: a theoretical basis for GIS. Proceedings of International Conference COSIT '97, Laurel Highlands. Springer, Berlin/Heidelberg/New York, pp1-13

Galton A (1997b) Space, time, and movement. In: Stock O (eds) Spatial and temporal reasoning. Kluwer Academic, Dordrecht, pp 321-352

Garson GD, Biggs RS (1992) Analytic mapping and geographic databases (Quantitative Applications in the Social Sciences 87). Sage Publications, Newbury Park

Gartner G (1999) Interaktive Karten im Internet. Ein Leitfaden. Springer, Berlin/Heidelberg/New York

Gascoigne J (1994) A la recherche du temps perdu (II): Incorporating time into GIS with respect to environmental modelling [http://www.geog.leeds.ac.uk/staff/m.blake/magis94/gisjg/exproc/ exproc.htm]

Gatrell A (1998) GIS and health (GISData 6). Taylor & Francis, London/Philadelphia

Gatrell A, Rowlingson B (1994) Spatial point process modelling in a GIS environment. Taylor & Francis, London

Gatrell AC (1991) Concepts of space and geographical data. In: Maguire DJ, Goodchild MF, Rhind DM (eds) Geographical Information Systems: principles and applications, Principles. Longman, Harlow, pp 119-134

Geffner H (1996) A qualitative model for temporal reasoning with incomplete information. Proceedings of the Thirteenth National Conference on Artificial Intelligence and Eighth Innovative Applications of Artificial Intelligence Conference, AAAI '96/IAAI '96, August 4-8, 1996, Portland, pp 1176-1181

Gegersen H, Jensen CS (1997) Temporal entity-relationship models-a survey. Technical Report 3. Time Center, Aalborg

George B, Haritsa JR (1997) Secure transaction processing in firm real-time database systems. Proceedings of the 1997 ACM SIGMOD International Conference on Mangement of Data, pp 462-473

George B, Haritsa JR, Seshadri S, Srinivasan V (1995) Index concurrency control in firm real-time database systems. Proceedings of the 21st International Conference on Very Large Databases, pp 134-145

Gerlach R (1991) Die kleinräumige Gliederung als Raumbezug. In: Schilcher M (ed) Geo-Informatik. Anwendungen, Erfahrungen, Tendenzen. Siemens Nixdorf, Berlin/München, pp 347-357

Gerstbach G (1989) Gegenwart und Zukunft geowissenschaftlicher Informationssysteme in Österreich. In: Dollinger F, Strobl J (eds) Beiträge zum GIS-Synopsium 5.-7. Juli 1989. Angewandte Geographische Informationstechnologien I (Salzburger Geographische Materialien). Institut für Geographie, Universität Salzburg, pp 27-34

Gertz M, Lipeck UW (1995) "Temporal" integrity constraints in temporal databases. In: Clifford J, Tuzhilin A (eds) Recent advances in temporal databases. Proceedings of the International Workshop on Temporal Databases, Zürich, Switzerland, 17-18 September 1995 (Workshops in Computing). Springer, Berlin/Heidelberg/New York, pp 77-92

Getis A (1964) Temporal land use pattern analysis with the use of nearest neighbour and quadrat methods. Annals of the Association of American Geographers 54:391-399

Getis A (1995) Spatial dependence and heterogeneity and proximal databases. In: Fotheringham S, Rogerson P (eds) Spatial Analysis and GIS. Taylor & Francis, London/Bristol, pp 105-120

Gillespie R, Davis WA (1981) Tree data structures for graphics and image processing. Proceedings of the 7th Conference of the Canadian Man-Computer Communications Socitey CMCCS '81/ACCHO '81, Waterloo, Canada, pp 155-162

Goh CH, Lu H, / Ooi BC, Tan KL (1995) Indexing temporal data using existing trees. Data & Knowledge 18:147-165

Gold CM (1992) The meaning of "Neighbour". In: Frank AE, Campari I, Formentini U (eds) Theories and methods of spatio-temporal reasoning in geographic space. International Conference GIS. From space to territory: Theories and methods of spatio-temporal reasoning, Pisa, Italy. Springer, Berlin/Heidelberg/New York, pp 220-235

Gold CM (1993) An outline of an event-driven spatial data structure for managing time-varying maps. Proceedings of the Canadian Conference on GIS, pp 880-888

Gold CM, Edwards G (1992) The Voronoi spatial model: two- and three-dimensional applications in image analysis. ITC Journal 1:11-19

Goldstein RC, Wagner C (1997) Database management with sequence trees and tokens. IEEE Transactions on Knowledge Representation and Reasoning 9:186-192

Gomboso J, Ghassemi F, Jakeman A (1993) Modelling ground water flow in the North stirling land conservation district, Western Australia. Proceedings of the International Congress on Modelling and Simulation. University of Western Australia, Perth, pp 735-740

Goodchild MF (1991a) Geographic Information Systems. Progress in Human Geography 15:194-200

Goodchild M (1991b) Just the facts. Political Quarterly Review 10:192-193

Goodchild MF (1992a) Geographical data modeling. Computer & Geosciences 18:401-408

Goodchild MF (1992b) Geographical information science. International Journal of Geographical Information Systems 6:31-45

Goodchild MF (1998) The geolibrary. In: Carver S (ed) Innovations in GIS 5. Selected Papers from the Fifth National Conference on GIS Research UK (GISRUK). Taylor & Francis, London/Bristol, pp 59-68

Goodchild MF (ed) (1989) The accuracy of spatial databases. Taylor & Francis, London

Goodchild MF et al. (eds) (1996) GIS and environmental modeling: progress and research issues. GIS World, Fort Collins

Goodchild MF et al. (eds) (1999) Interoperating Geographic Information Systems (Kluwer International Series in Engineering and Computer Science 495). Kluwer Academic Publishers, Boston

Goodchild M, Haining R, Wise S (1992) Integrating GIS and spatial data analysis: problems and possibilities. International Journal of Geographical Information Systems 6:407-423

Goodchild MF, Anselin L, Deichmann U (1993) A framework for the areal interpolation of socioeconomic data. Environment and Planning 25:383-397

Goodchild MF, Lam SN (1980) Areal interpolation: A variant of the traditional spatial problem. Geo-Processing 1:297-312

Goodchild MF, Parks BO, Steyaert LT (eds) (1993) Environmental modelling with GIS. Oxford University Press, Oxford

Goodno BJ, Wright JR (eds) (1992) Computing in civil engineering and Geographic Information Systems. Symposium: Proceedings of the Eighth Conference Dallas, Texas June 7-9, 1992. American Society of Civil Engineers, New York

Göpfert W (1991) Raumbezogene Informationssysteme. Grundlagen der integrierten Verarbeitung von Punkt-, Vektor- und Rasterdaten. Anwendungen in Kartographie, Fernerkundung und Umweltplanung, 2nd ed. Wichmann, Karlsruhe

Goralwalla IA, Leontiev Y, Özsu MT, Szafron D (1997) Modeling temporal primitives: back to basics. Proceedings of the Sixth International Conference on Information and Knowledge Management CIKM'97, Las Vegas, November 10-14, 1997, pp 24-31

Goralwalla IA, Özsu MT, Szafron D (1997) A framework for temporal data models: exploiting object-oriented technology. Paper presented at the 13th International Conference and Exhibition on Technology of Object-Oriented Languages and Systems (TOOLS '98), 3-7 August, 1998, Santa Barbara

Goralwalla IA, Szafron D, Özsu MT, Peters RJ (1997) Managing schema evolution using a temporal object model. In: Embley DW, Goldstein RC (eds) Conceptual Modeling-ER '97, 16th International Conference on Conceptual Modeling, Los Angeles, November 3-5, 1997, Proceedings (LNCS 1331), pp 71-84

Gould P (1985) The geographer at work. Routledge & Kegan, London/Boston/Henley

Gould SJ (1987) Time's arrow time's cycle. Myth and metaphor in the discovery of geological time. Harvard University Press, Cambridge

Grandi F, Scalas MR (1997) Extending temporal database concepts to the world wide web. Technical Report CSITE 004-97. University of Bologna, Bologna

Green D, Rix D (eds) (1994) The AGI sourcebook for Geogaphic Information Systems. Association for Geographic Information, London

Green D, Yao F (1986) Finite resolution computional geometry. In: Institute of Eletrical and Electronics Engineers (ed) Proceedings of the 27th IEEE Symposium on Foundations of Computer Science. IEEE, New York, pp 143-152

Greene RW (2000) GIS and public policy. ESRI Press, Redlands

Gregersen H, Jensen CS, Mark L (1997) Evaluating temporally extended ER-models. Paper presented at the 2nd CAiSE97 International Workshop on Evaluation of Modeling Methods in Systems Analysis and Design, Barcelona, 1997

Gregory D (1994) Geographical imaginations. Blackwell, Oxford

Gregory I, Bennett C (1997) Local history & Geographical Information Systems. Local History Magazine 62:19-22

Gregory I, Gilham V (1998) Historical GIS: a framework for mapping the past. Bulletin of the Society of Cartographers 31:9-16

Gregory I, Gilham V, Southall H (1998) Looking forward to the past. Mapping awareness 12:26-29

Gregory IN, Southall HR (1998) Putting the past in its place: the Great Britain historical GIS. In: Carver S (ed) Innovations in GIS 5. Selected Papers from the Fifth National Conference on GIS Research UK (GISRUK). Taylor & Francis, London/Bristol, pp 210-221

Grelot JP, Chambon P (1984) Updating a landuse inventory. In: Douglas DH (ed) Proceedings of AutoCarto 6. University of Toronto Press, Toronto, pp 44-51

Greve C, Kelmelis J, Fegeas R, Guptill S, Mouat N (1993) Investigating U.S. geological survey needs for the management of temporal GIS data. Photogrammetric Engineering & Remote Sensing 59:1503-1507

Greve K (1990) Konzeptionelle Überlegungen zum Einsatz eines Rauminformationssystems in der Historischen Stadtforschung. Das Beispiel Altona 1803. In: Dollinger F, Strobl J (eds) Beiträge zum GIS-Symposium 9.-11. Juli 1990. Angewandte Geographische Informationstechnologien II (Salzburger Geographische Materialien). Institut für Geographie, Universität Salzburg, pp 171-180

Greve K (1995) Metainformationssysteme und Umweltinformation. Thesen zum Stand der Forschung und Entwicklung. In: Dollinger F, Strobl J (eds) Beiträge zum GIS-Symposium 5.-7. Juli 1995. Beiträge zur Geographischen Informationswissenschaft VII (Salzburger Geographische Materialien). Institut für Geographie, Universität Salzburg, pp 207-212

Greve K (1997) Objekte der Metadatenverarbeitung im Umweltbereich. In: De Lange N (ed) Metadaten. Konzepte und Anwendungen in der raumbezogenen Datenverarbeitung. Osnabrücker Studien zur Geographie-Materialien. Selbstverlag, Osnabrück, pp 27-42

Griffith DA (1993) Advanced spatial statistics for analysing and visualizing geo-referenced data. International Journal of Geographical Information Systems 7:107-123

Griffith DA (1993) Which spatial statistics techniques should be converted to GIS functions? In: Fischer MM, Nijnkamp P (eds) Geographic Information Systems, spatial modelling and policy evaluation. Springer, Berlin/Heidelberg/New York, pp 103-114

Grimshaw DJ (2000) Bringing Geographical Information Systems into business, 2nd ed. John Wiley & Sons, New York/Weinheim

Gripaios PA, Mangles T (1994) An analysis of European super regions. Regional Studies 27:745-750

Grossman WD, Eberhardt S (1992) Geographical Information Systems and dynamic modelling: potentials of a new approach. Annals of Regional Science 26:53-66

Grünreich D (1992) Aufbau von Geo-Informationssystemen im Umweltschutz mit Hilfe von ATKIS. In: Günther O, Schulz KP, Seggelke J (eds) Umweltanwendungen geographischer Informationssysteme. C.F. Müller, Karlsruhe, pp 3-14

Guesgen HW (1989) Spatial reasoning based on Allen's temporal logic. International Computer Science Institute, Berkeley

Günther O (1992) Datenbanktechniken zur Verwaltung räumlicher Daten. In: Günther O, Schulz KP, Seggelke J (eds) Umweltanwendungen Geographischer Informationssysteme. Wichmann, Karlsruhe, pp 15-23

Günther O (1998) Environmental Information Systems. Springer, Berlin/Heidelberg/New York

Günther O, Lamberts J (1992a) Object-oriented techniques for the management of geographic and environmental data. Forschungsinstitut für anwendungsorientierte Wissensverarbeitung an der Universität Ulm FAW-TR-92023. FAW, Ulm

Günther O, Lamberts J (1992b) Objektorientierte Techniken zur Verwaltung von Geodaten. In: Günther O, Riekert WF (eds) Wissensbasierte Methoden zur Fernerkundung der Umwelt. Wichmann, Karlsruhe, pp 55-90

Günther O, Lamberts J (1994) Object-oriented techniques for the management of geographic and environmental data. The Computer Journal 37:16-25

Günther O, Riekert WF (1993) The design of GODOT: an object-oriented Geographic Information System. IEEE Data Engineering Bulletin 16:4-9

Guptill SC (1992) Multiple representations of geographic entities through space and time. Proceedings of the Fourth International Symposium on Spatial Data Handling, pp 859-868

Gurnell AM, Montgomery DR (eds) (2000) Hydrological applications of GIS (Advances in hydrological processes). John Wiley & Sons, Chichester/Weinheim

Güssefeldt J (1997) Grundsätzliche Überlegungen zu Regionalisierungsmodellen. Geographische Zeitschrift 85:1-19

Güting R (1989) GRAL: an extensible relational database system for geometric applications. In: Apers PMG, Wiederhold G (eds) Proceedings of the Fifteenth International Conference on Very Large Data Bases, August 22-25, 1989, Amsterdam, pp 33-44

Güting RH, De Ridder T, Schneider M (1995) Implementation of the ROSE algebra: efficient algorithms for realm-based spatial data types. In: Egenhofer MJ, Herring JR (eds) Advances in Spatial Databases, Fourth International Symposium, SSD'95, Portland, Maine, August 6-9, 1995, Proceedings (LNCS 951). Springer, Berlin/Heidelberg/New York, pp 216-239

Güting RH, Schneider M (1993) Realms: a foundation for spatial data types in database systems. In: Abel D, Ooi BC (eds) Advances in spatial databases, Third International Symposium, SSD 93 (LNCS 692). Springer, Berlin/Heidelberg/New York, pp 14-35

Guttmann A (1984) R-trees: a dynamic index structure for spatial searching. Proceedings of SIGMOD. SIGMOD, Boston, MA, pp 47-57

Hägerstrand T (1952) The propagation of innovation waves. Lund Studies in Geography B, Human Geography 4:3-19

Hägerstrand T (1970) What about people in regional science? Regional Science Association Papers 24:7-21
Haggett P, Cliff AD, Frey A (1977) Locational analysis in human geography. Edward Arnold, London
Haines-Young R (1989) Modelling geographic knowledge. In: MacMillan B (ed) Remodelling Geography. Blackwell, Oxford, pp 22-39
Haining RP (1981) Spatial and temporal analysis: spatial modelling. In: Wrigley N, Bennett RJ (eds) Quantitative geography: a British view. Routledge & Kegan, London/Boston/Henley, pp 86-91
Hake G, Grünreich D (1994) Kartographie. De Gruyter, Berlin/New York
Halls J, Cowen D, Jensen J (1994) Predictive spatio-temporal modeling in GIS. In: Waugh TC, Healey RG (eds) Advances in GIS research. Proceedings of the 6th International Symposium on Spatial Data Handling, Edinburgh. Taylor & Francis, London, pp 431-447
Halls P, Miller P (1995) Moving GIS into the fourth dimension. Or the case for todes. Proceedings of the GIS Research UK 1995 Conference, Newcastle upon Tyne, pp 41-43
Halls P, Miller P (1996) Of todes and worms: an experiment in bringing time to ArcInfo [http://www.arcdataonline.com/library/userconf/europroc96/PAPERS/PN35/PN35F.HTM]
Hammer M, McLeod D (1981) Database description with SDM: a semantic database model. ACM Transactions on Database Systems 6:351-386
Hamre T (1994) An object-oriented conceptual model for measured and derived data varying 3D space and time. In: Waugh TC, Healey RG (eds) Advances in GIS research. Proceedings, 6th International Symposium on Spatial Data Handling, Edinburgh. Taylor & Francis, London, pp 868-881
Hanna KC (2000) GIS for landscape architects. ESRI Press, Redlands
Harder C (1997) ArcView GIS means business. ESRI Press, Redlands
Harder C (1999) Enterprise GIS for energy companies. ESRI Press, Redlands
Haritsa JR, Seshadri S (1996) Ral-time index concurrency control. SIGMOD Record 25:13-17
Harley JB (1989) Deconstructing the map. In: Barnes T, Duncan J (eds) Writing world: discourse, text and metaphor in the representation of landscape. Routledge, London, pp 231-247
Harley JB (1990) Cartography, ethics and social theory. Cartographica 27:1-23
Harmon D, Limp WF (1998) Inside Geomedia. OnWord Press, Santa Fe
Harris W, Gray WA (1995) Using temporal constructs in temporal databases. In: Clifford J, Tuzhilin A (eds) Recent advances in temporal databases. Proceedings of the International Workshop on Temporal Databases, Zürich, Switzerland, 17-18 September 1995 (Workshops in Computing). Springer, Berlin/Heidelberg/New York pp 133-152
Harrower M (1999) A conceptual framework for mapping space-time [http://www.geog.psu.edu/~harrower/pdf/cube.zip]
Hartman R (1997) Focus on GIS component software: featuring ESRI's mapobjects. OnWord Press, Santa Monica
Harts JJ, Van Lammeren R (1997) Analysefuncties in GIS. In: Hendriks P, Ottens H (eds) Geografische Informatie Systemen in ruimtelijk onderzoek. Van Gorcum, Assen, pp 115-144
Harvey F, Chrisman N (1998) Boundary objects and the social construction of GIS technology. Environment and Planning A 30:1683-1694
Hasting H, Perkelney R, Monticciolo R, Vun-Kannon D, Del-Monte D (1982) Time scales, persistence and patchiness. BioSystems 15:281-289
Hayes BR (1999) Modeling the environment with ArcView 3.1. OnWord Press, Santa Fe
Hazelton NWJ (1991a) Extending GIS to include dynamic modelling. Proceedings of the Third Colloquium of the Spatial Information Research Centre, University of Otago, Dunedin, New Zealand, 27th-28th May, pp 73-82
Hazelton NWJ (1991b) Integrating time, dynamic modelling and Geographical Information Systems: development of four-dimensional GIS. Unpublished Ph.D. thesis, Department of Surveying and Land Information, University of Melbourne
Hazelton NWJ (1991c) Using GIS with non-point-source and distributed models: the 'how' and 'why' of an information opportunity. Proceedings of the AURISA '91 Conference, Wellington, New Zealand, November. AURISA, Wellington, pp 298-314
Hazelton NWJ (1992a) Beyond the 2-D Map: a new metaphor for multi-temporal 4-D GIS. Proceedings GIS/LIS'92. San Jose, pp 303-313
Hazelton NWJ (1992b) Developments in spatio-temporal GIS. Paper presented at the „Geographical Information Sciences Research in Victoria and Tasmania“ Conference, Ballarat

University College, 24th-25th September, 1992 [http://geomatics.eng.ohio-state.edu/ Papers/Ballarat_92/Ballarat_92.html]
Hazelton NWJ (1993) Modeling in temporal GIS. MicroStation Manager 3:58-60
Hazelton NWJ (1997) Temporal aspects of maps and mapping. Some implications. Surveying and Land Information Systems 57:42-54
Hazelton NWJ (1998) Some operational requirements for a multi-temporal 4-D GIS. In: Egenhofer MJ, Golledge RG (eds) Spatial and temporal reasoning in Geographic Informations Systems. Oxford University Press, New York/Oxford, pp 63-73
Hazelton NWJ, Bennett L. Masel J (1992) Topological structures for 4-dimensional Geographic Information Systems. computers, Environment and Urban Systems 16:227-237
Hazelton NWJ, Leahy FJ, Williamson IP (1990a) Incorporating dynamic modelling into GIS. Proceedings of the AURISA '90/URPIS 18 Conference, Canberra, pp 77-84
Hazelton NWJ, Leahy FJ, Williamson IP (1990b) On the design of temporally referenced 3D Geographical Information Systems: development of 4-dimensional GIS. Proceedings of GIS/LIS '90, Anaheim. ACSM, Bethesda, pp 357-372
Hazelton NWJ, Leahy FJ, Williamson IP (1993) Integrating dynamic modelling and Geographical Information Systems. Journal of URISA 4:47-58
Hazelton NWJ, Pigot S (1992) Basic concepts in 3-D and 4-D Geographic Information Systems and their application in the earth sciences. Proceedings of the Conference on the Geological Applications of Geographic Information Systems, pp 9-19
Healey CG (1997) On the use of perceptual cues and data mining for effective visualization of scientific datasets. Proceedings of the 24th Conference on Graphics Interface (GI-98), June 18-20 1998. Morgan Kaufmann, San Mateo, pp 177-184
Healey RG (1991) Database management systems. In: Maguire DJ et al. (eds) Geographical Information Systems, principles and technical issues. John Wiley & Sons, New York, pp 251-267
Healey RG, Dowers S, Minetar MJ (eds) (1998) Parallel processing and GIS. Taylor & Francis, London/Bristol
Hearnshaw HM, Unwin DJ (1994) Visualization in GIS. John Wiley & Sons, Chichester
Heit M et al. (ed) (1991) GIS applications in natural resources. GIS World, Fort Collins
Heit M et al. (ed) (1996) GIS applications in natural resources 2. GIS World, Fort Collins
Helgerson R (1989) Dismantling to build. Cartographica 26:98-101
Hellwig, F (1999): Ein objektorientiertes Datenmodell für ein zeitintegratives GIS. Unpublished thesis. Department of Geography University of Mannheim. Mannheim [http://members.aol.com/FHTGIS/tgis.html]
Helokunnas T (1994) Object-oriented geographic data management. EGIS-Foundation (ed) 5th European Conference and Exhibition on GIS. EGIS, Paris, pp 1194-1203
Hendriks P, Ottens H (1997a) Kennismaking met GIS. In: Hendriks P, Ottens H (eds) Geografische informatie systemen in ruimtelijk onderzoek. Van Gorcum, Assen, pp 1-20
Hendriks P, Ottens H (1997b) De toekomst van GIS. In: Hendriks P, Ottens H (eds) Geografische informatie systemen in ruimtelijk onderzoek. Van Gorcum, Assen, pp 195-198
Hendriks P, Ottens H (eds) (1997c) Geografische informatie systemen in ruimtelijk onderzoek. Van Gorcum, Assen
Hepple LW (1981) Spatial and temporal analysis: time series analysis. In: Wrigley N, Bennett RJ (eds) Quantitative geography: a British view. Routledge & Kegan, London/Boston/Henley, pp 92-96
Heres L (2000a) Hodochronologics: history and time in the National Road Database. In: Heres L (ed): Time in GIS: issues in spatio-temporal modelling (Publications on geodesy 47). Nederlandse Commissie voor Geodesie, Delft, pp 46-56
Heres L (2000b) Time in GIS. issues in spatio-temporal modelling. In: Heres L (ed) Time in GIS: issues in spatio-temporal modelling (Publications on geodesy 47). Nederlandse Commissie voor Geodesie, Delft, pp1-2
Hermes T (1998) GIS-basierte Konzepte zum Geo-Marketing unter besonderer Berücksichtigung der Vertriebsoptimierung. Geoinformatik online 1 [http://gio.uni-muenster.de/beiträge/ausg1_98/hermes/start.htm]
Hermosilla LH (1994) A unified approach for developing a temporal GIS with database and reasoning capabilities. Proceedings of the Fifth European Conference and Exhibition on Geographic Information Systems, EGIS '94. EGIS Foundation, Utrecht, pp 122-131 [http://www.odyssey.ursus.maine.edu/gisweb/spatdb/egis/eg94014.html]

Hermsdörfer D, Widmer M (1998) Mit ArcView ins Data Warehouse SIS. Mengenbezogene Raumanalyse für flexibles Informationsmanagement. ESRI Arc Aktuell 1:10-11
Herring JR (1987) TIGRIS: Topologically integrated geographic information system. Proceedings of AutoCarto 8, pp 282-291
Herring JR (1991) Using spline functions to represent distributed attributes. Proceedings of AutoCarto 10, pp 46-58
Herring JR (1992) TIGRIS: a data model for an object-oriented geographic information system. Computers & Geosciences 18:443-452
Herrmann C, Asche H (1998) Animierte thematische Karten zur Schweizer Bevölkerungsstatistik. Geographica Helvetica 53:17-19
Hertig L, Maus O (1998) Geomarketing bei Banken und Sparkassen. Geographische Informationssysteme 11:13-16
Heuer A (1992) Objektorientierte Datenbanken. Addison-Wesley, Bonn
Heuvelink GBM et al. (1998) Error propagation in environmental modelling with GIS (Research Monographs in GIS). Taylor & Francis, London
Heywood I (1990) GIS in the social sciences. Environment & Planning A 22:849-854
Heywood I, Cornelius S, Carver S (1998) An introduction to Geographical Information Systems. Longman, Harlow
Hing SH, Kim MH (1997) Resolving data conflicts with multiple versions and precedence relationships in real-time databases. Information Processing Letters 61:149-156
Hintz R, Onsrud H (1990) A methodology for upgrading real property boundary information in a GIS using a temporally efficient automated survey measurement management system. Department of Surveying Engineering, University of Maine, Orono
Hirtle SC, Ghiselli-Crippa T, Spring MB (1995) The cognitive structure of space: an analysis of temporal sequences. In: Frank AU, Kuhn W (eds) Spatial information theory-a theoretical basis for GIS. Proceedings COSIT '95 (LNCS 988). Springer, Berlin/Heidelberg/New York, pp 177-189
Hjesvold R, Midtstraum R, Sandsta O (1995) A temporal foundation of video databases. In: Clifford J, Tuzhilin A (eds) Recent advances in temporal databases. Proceedings of the International Workshop on Temporal Databases, Zürich, Switzerland, 17-18 September 1995 (Workshops in Computing). Springer, Berlin/Heidelberg/New York, pp 295-314
Hofmann-Wellenhof B, Lichtenegger H, Collins J (1997) Global positioning system. Theory and practice, 4th ed. Springer, Wien
Hohl P (ed) (1998) GIS data conversion: strategies, techniques, and management. OnWord Press, Santa Fe
Hohl P, Mayo B (1997) Arcview Gis exercise book. OnWord Press, Santa Fe
Hoinkes R, Mitchell R (1994) Playing with time: continuous temporal mapping strategies for interactive environments [http://www.clr.toronto.edu:1080/COLLAB/CCA/ContTemp/ gispaper.rtf.binary.html]
Holtz WB (1997) The CAD rating guide: a tool for the evaluation of computer-aided design systems, including FEM, GIS, and animation systems, 5th ed. Penn Well Publishers, Tulsa
Hornsby K, Egenhofer M (2000) Identity-based change: a foundation for spatio-temporal knowledge representation. International Journal of Geographical Information Science 14:207-224
Huang J, Gruenwald L (1996) Impact of timing constraints on real-time database recovery. Paper presented at the Fifth International Conference on Information and Knowledge Management, Workshop on databases: active and real-time, Rockyville, MD, November 12-16, 1996
Huber M, Schneider D (1999) Spatial data standards in view of models of space and the functions operating on them. Computers & Geosciences 25:25-38
Hull R, King R (1987) Semantic database modeling: survey, applications, and research issues. ACM Computing Surveys 19:201-260
Hunter GJ (1986) Non-current data and land information systems. Unpublished Master thesis, Department of Surveying and Land Information, University of Melbourne
Hunter GJ (1988) Non current data and Geographical Information Systems: a case for data retention. International Journal of Geographical Information Systems 2:281-286
Hunter GJ (1992) Selection criteria for archiving non-current data in land information systems. CISM Journal 46:255-264
Hunter G, Williamson I (1990) The development of a historical digital cadastral database. International Journal of Geographical Information Systems 4:169-179
Hutchinson S, Daniel L (1997) Inside Arcview Gis, 2nd ed. OnWord Press, Santa Fe

Huxhold WE (1991) An introduction to urban Geographic Information Systems. Oxford University Press, New York/Oxford

Huxhold WE, Levinsohn AG (1995) Mananging Geographic Information System projects. Oxford University Press, New York/Oxford

Illert A (1997) MERGIN: Kooperation der Amtlichen Kartographie in Europa. In: Grünreich D (ed) GIS und Kartographie im multimedialen Umfeld. Grundlagen, Anwendungen und Entwicklungstrends (Kartographische Schriften). Kirschbaum, Bonn, pp 78-82

Inmon WH (1986) Information systems architecture: a system developer's primer. Prentice Hall, Englewood Cliffs

Isard W (1970) On notions and models of time. Papers in Regional Science 25:7-32

Itami RM (1988) Cellular automatons as a framework for dynamic simulations in Geographic Infomation Services. Proceedings of GIS/LIS '88, San Antonio, pp 590-597

Jäschke UU, Müller M (1999) Zur Problematik der Anpassung von historischen Karten an moderne Koordinatensysteme. In: Ebeling D (ed) Historisch-thematische Kartographie: Konzepte–Methoden–Anwendungen. Verlag für Regionalgeschichte, Bielefeld, pp 150-166

Jenks GF (1963) Generalization in statistical mapping. Annuals of the Association of American Geographers 53:15-26

Jensen CS, Snodgrass RT (1996a) Semantics of time-varying attributes and their use for temporal database design. In: Papazoglou MP (ed) OOER'95: object-oriented and entity-relationship modelling. 14th International Conference, Gold Coast, Australia, December 12-15, 1995. Proceedings. (LNCS 1021). Springer, Berlin/Heidelberg/New York, pp 366-377

Jensen CS, Snodgrass RT (1996b) Semantics of time-varying information. Information Systems 21:311-352

Jensen CS, Snodgrass RT (1997a) Semantics of time-varying attributes and their use for temporal database design. Technical Report 1. Time Center, Aalborg

Jensen CS, Snodgrass RT (1997b) Temporal data management. Technical Report 17. Time Center, Aalborg

Jensen CS, Snodgrass RT, Soo MD (1992) Extending normal forms to temporal relations. Technical Report 92-17. University of Arizona, Department of Computer Science

Jensen CS, Soo MD, Snodgrass RT (1992) Unification of temporal data models [ftp://ftp.cs.arizona.edu/reports/1992/TR92-15.ps]

Jhingran A (1996) Why commercial databases systems are not real-time systems. Paper presented at the Fifth International Conference on Information and Knowledge Management, Workshop on databases: active and real-time, Rockyville, MD, November 12-16, 1996

Jixin M, Knight B (1996) Building temporal constraints into knowledge bases for process control-an examination. Engineering Applications of Artificial Intelligence 9:95-96

Joao EM (1998) Causes and consequences of map generalisation (Research Monographs in Geographic Information Systems). Taylor & Francis, London

Joerin F, Claramunt C (1994) Integrating the time component in a GIS: an application to assess flooding impacts on agriculture. In: Harts JJ, Ottens HFL, Scholten HJ (eds) EGIS '94 European Conference and Exhibition on GIS Proceedings, Paris, pp 524-532

Johnson AI et al. (1992) Geographic Information Systems (GIS and Mapping-Practices and Standards). American Society for Testing and Materials, Philadelphia, PA

Johnson D, Gluck, M (1997) Geographic information retrieval and the World Wide Web: A match made in electronic space. Cartographic Perspectives 26:13-26

Johnson I (1996) Understanding MapInfo: a structured guide. Archaeological Computing Laboratory, Sydney

Johnston CA (1998) Geographic Information Systems in ecology. Blackwell, Oxford

Jones CB (1989) Data structures for three-dimensional spatial information systems in geology. International Journal of Geographical Information Systems 3:15-31

Jones CB (1997) Geographical Information Systems and computer cartography. Longman, Harlow

Jones MB (1996) Consumer real-time systems. ACM Computer Surveys 28:180

Jukic N, Vrbsky SV (1996) Feasibility of aggregates in temporal constraint queries. Information Systems 21:595-614

Kaikova H, Terziyan V (1997) Temporal knowledge acquisition from multiple experters. Paper presented at the Third International Workshop on Next Generation Information Technologies and Systems (NGITS '97), Neve Ilan, Israel

Kaiser A (1996) Zeitbezogene Datenbanksysteme-eine Bestandsaufnahme und ausgewählte Problemstellungen. Proceedings of the 5. Internationales Symposium für Informationswissenschaft ISI '96, pp 143-155

Kaiser A (1997) Umsetzung zeitbezogener Konzepte in relationalen Datenbanksystemen. In: Kracke U (ed) Datenbank-Management. Interest-Verlag, Augsburg, pp 7-8

Kaiser A (1998) Überlegungen zur Entitätsintegrität in temporalen relationalen Datenbanksystemen. Proceedings of the Fourth ZOBIS-Workshop, pp 74-78

Kaiser A, Panny W (1996) Die Abbildung zeitbezogener Daten in Datenbanksystemen mit Standard-SQL. Proceedings of the ZOBIS-Workshop, Cottbus, pp 63-66

Kaiser A, Panny W (1997) Möglichkeiten der Abbildung zeitbezogener Daten in Datenbanksystemen mit SQL. Arbeitsberichte zum Tätigkeitsfeld Informationsverarbeitung und Informationswirtschaft, Wirtschaftsuniversität Wien. Wien

Kampke T (1994) Storing and retrieving changes in a sequence of Polygons. International Journal of Geographica Information Systems 8:493-513

Katz RH, Anwarrudin M, Chang E (1986) Organizing a design database across time. In: Brodie M, Mylopoulos J (eds) Large scale knowledge-based systems. Springer, Berlin/Heidelberg/New York, pp 287-296

Katz RH, Chang E, Bhateja R (1986) Version modeling concepts for computer-aided design databases. Proceedings of the SIGMOD '86 Conference. ACM, New York, pp 379-386

Keates JS (1982) Understanding maps. Longman, Harlow

Keitel A, Müller M (1995) Die Integration von Sachdaten, Geodaten und Metadaten im Umweltinformationssystem Baden-Württemberg. In: Kremers H, Pillmann W (eds) Raum und Zeit in Umweltinformationssystemen. 9th International Symposium on Computer Science for Environmental Protection CSEP '95 (Umwelt-Informatik aktuell). Metropolis-Verlag, Marburg, pp 400-407

Keller PC, Waters NM (1991) Mapping software for microcomputers. In: Fraser Taylor DR (ed) Geographic Information Systems. Modern Cartography. Pergamon Press, Oxford, pp 97-128

Keller SF (1999) Modelling and sharing geographic data with INTERLIS. Computers & Geosciences 25:49-59

Kelmelis J (1991) Time and space in geographic information: toward a 4-dimensional spatio-temporal data model. Unpublished Ph.D. thesis, Department of Geography, The Pennsylvania State University

Kelmelis JA (1998) Process dynamics, temporal extent, and causal propagation as the basis for linking space and time. In: Egenhofer MJ, Golledge RG (eds) Spatial and temporal reasoning in Geographic Informations Systems. Oxford University Press, New York/Oxford, pp 94-103

Kemp KK (1993) Environmental modeling with GIS: a strategy for dealing with spatial continuity. NCGIA Technical Report 93-3, Santa Barbara

Kemp Z, Kowalczyk A (1994) Incorporating the temporal dimension in a GIS. In: Worboys MF (ed) Innovations in GIS. Taylor & Francis, London, pp 89-103

Kemper A, Eickler A (1996) Datenbanksysteme. Oldenbourg, München

Kendall MG, Yule GU (1950) An introduction to the theory of statistics. Grifflin, London

Kennedy M (1996) The global positioning system and GIS: an introduction. Ann Arbor Press, Chelsea

Keogh E, Smyth P (1997) A probablistic approach to fast pattern matching in time series databases. In: Heckerman D, Mannila H, Pregibon D (eds) Proceedings of the Third International Conference on Knowledge Discovery and Data Mining (KDD-97), Newport Beach, August 14-17, 1997, pp 24-30

Keravnou ET (1996) Engineering time in medical knowledge-based systems through time-axes and time-objects. Paper presented at the Third International Workshop on Temporal Representation and Reasoning TIME '96

Kick U (1996) GIS und Logistik. Lektion 13: Logistikanwedungen auf der Basis von Straßennetzdaten. In: Bumann E, Bachhuber R, Schaller J (eds) ArcView. GIS-Arbeitsbuch. Wichmann, Heidelberg, pp 219-230

Kienast F, Frank C, Leu R (1991) Analyse raum-zeitlicher Daten mit einem Geographischen Informationssystem. Technical Report, Berichte der eidgenössischen Forschungsanstalt für Wald, Schnee und Landschaft

Kilchenmann A (ed) (1992) Technologie Geographischer Informationssysteme. Springer, Berlin/Heidelberg/New York

Kilchenmann A, Schwart von Raumer HG (eds) (1999) GIS in der Stadtentwicklung: Methodik und Fallbeispiele. Springer, Berlin/Heidelberg/New York

Kim J, Yoo H, Lee Y (1990) Design and implementation of a temporal query language with abstract time. Information Systems 15:349-357

Kim JS, Kim MH (1997) On effective data clustering in bitemporal databases. Proceedings of the Fourth International Workshop on Temporal Representation and Reasoning, TIME '97, May 10-11, 1997, Daytona Beach

Kirner TG, Davis AM (1996) Requirements specification of real-time systems: temporal parameters and timing-constraints. Information and Software Technology 38:735-741

Kirousis LM, Spirakis PG, Tsigas P (1996) Simple atomic snapshots: a linear complexity solution with unbounded time-stamps. Information Processing Letters 58:47-53

Klemmer W, Spranz R (1997) GIS-Projektplanung und Projektmanagement: Theorie und Praxis. Selbstverlag, Bonn

Knight B, Jixin M (1997) Temporal management using relative time in knowledge-based process control. Engineering Applications of Artificial Intelligence 10:269-280

Knolmayer G, Buchberger T (1997) Maintaining temporal integrity of world wide web pages. Chapman & Hall, London

Knolmayer G, Myrach T (1996) Zur Abbildung zeitbezogener Daten in betrieblicher Informationssystemen. Wirtschaftsinformatik 38:63-74

Knolmayer G, Myrach T (1997) Die Berücksichtigung fehlerhafter Daten durch temporale Datenhaltung. In: Fischer TR, Hömberg R (eds) Jahresabschluss und Jahresabschluss prüfung, Probleme, Perspecktiven, internationale Einflüsse. IDW, Düsseldorf, pp 863-905

Knox E (1964) The detection of space-time interactions. Applied Statistics 13:25-29

Koch A (1998) GIS und Systemmodelle zur Visualisierung und Simulation dynamischen Systemverhaltens. In: Strobl J, Dollinger F (eds) Angewandte Geographische Informationsverarbeitung. Beiträge zur AGIT-Symposium, Salzburg 1998. Wichmann, Heidelberg [http://www.geo.sbg.ac.at/agit/pqpers98/koch/Koch.html]

Köck H (1997) Die Rolle des Raumes als zu erklärender und als erklärender Faktor. Zur Klärung einer methodologischen Grundrelation in der Geographie. Geographica Helvetica 52:89-96

Kohli M (1996) Real-time databases in telecommunications. Paper presented at the Fifth International Conference on Information and Knowledge Management, Workshop on databases: active and real-time, Rockyville, MD, November 12-16, 1996

Kollarits S (1994) Mobilität und GIS: Probleme der Repräsentation und Analyse. In: Dollinger F, Strobl J (eds) Beiträge zum GIS-Synopsium 6.- 8. Juli 1994. Angewandte Geographische Informationstechnologien VI (Salzburger Geographische Materialien). Institut für Geographie, Universität Salzburg, pp 371-384

Kollarits S, Heuegger M, Uschnigg M (1998) Das Modifiable Areal Unit Problem: Diskussion, Implementation und empirische Effekte. In: Strobl J, Dollinger F (eds) Angewandte Geographische Informationsverarbeitung. Beiträge zum AGIT-Symposium Salzburg '98. Wichmann, Heidelberg, pp 170-179

Kollarits S, Riedl L (1992) Adressen, Straßengraph und Baublocks: grundlegende Beziehungen und deren praktische Umsetzung. In: Dollinger F, Strobl J (eds) Beiträge zum GIS-Synopsium 8.-10.Juli 1992. Angewandte Geographische Informationstechnologien IV (Salzburger Geographische Materialien). Institut für Geographie, Universität Salzburg, pp 63-78

Kollios G, Gunopulos D, Tsotras VJ (1999) Nearest neighbor queries in a mobile environment. In: Böhlen MH, Jensen CS, Scholl MO (eds) Spatio-temporal database management. Proceedings of the International Workshop STDBM'99 Edinburgh, September 10-11, 1999. Springer, Berlin/Heidelberg/New York, pp 119-134

Kolovson C (1990) Indexing techniques for multi-dimensional spatial data and historical data in database management systems. Electronics Reseach Laboratory. Technical Report UCB/ERL M90/105, University of Berkeley

Kolovson C (1993) Indexing techniques for historical databases. In: Tansel A, Clifford J, Gadia SK, Jajodia S, Segev A, Snodgrass R (eds) Temporal databases: theory, design, and implementation. Benjamin Cummings, Redwood City, pp 418-432

Kolovson C, Stonebraker M (1991) Segment indexes: dynamic techniques for multi-dimensional interval data. In: Clifford J, King R (eds) Proceedings of ACM SIGMOD 1991, pp 138-147

Konana P, Gupta A, Whinston AB (1996) Research issues in real-time DBMS in the context of electronic commerce. Paper presented at the Fifth International Conference on Information and Knowledge Management, Workshop on databases: active and real-time, Rockyville, MD, November 12-16, 1996

Koob G (1996) Large scale distributed real-time systems. ACM Computer Surveys 28:181

Korn F, Jagadish HV, Faloutsos C (1997) Efficiently supporting ad hoc queries in large datasets of time sequences. In: Peckham, Joan (ed): SIGMOD 1997, Proceedings ACM SIGMOD International Conference on Management of Data, May 13-15, 1997, Tucson

Korte B (1997) The GIS Book: understanding the value and implementation of Geographic Information Systems, 4th ed. OnWord Press, Santa Fe

Korth HF (1996) Database system architectures for time-constrained applications. Paper presented at the Fifth International Conference on Information and Knowledge Management, Workshop on databases: active and real-time, Rockyville, MD, November 12-16, 1996

Korzybski A (1958) Science and sanity: an introduction to non-Aristotlean systems and semantics, 4th ed. Institute of General Semantics, Lakeville

Kösters G (1997) Requirements engineering für GIS-Applikationen. Shaker, Aachen

Koubarakis M (1993) Representation and querying in temporal databases: the power of temporal constraints. Proceedings of the 9th International Conference on Data Engineering, pp 327-334

Koubarakis M (1995) Databases and temporal constraints: semantics and complexity. In: Clifford J, Tuzhilin A (eds) Recent advances in temporal databases. Proceedings of the International Workshop on Temporal Databases, Zürich, Switzerland, 17-18 September 1995 (Workshops in Computing). Springer, Berlin/Heidelberg/New York, pp 93-109

Koubarakis M (1997a) From local to global consistency in temporal constraint networks. Theoretical Computer Science 73:89-112

Koubarakis M (1997b) The complexity of query evaluation in indefinit temporal constraint databases. Theoretical Computer Science 171:25-60

Kraak MJ (2000) Visualisation of the time dimension. In: Heres L (ed): Time in GIS: issues in spatio-temporal modelling (Publications on geodesy 47). Nederlandse Commissie voor Geodesic, Delft, pp 27-35

Kraak MJ (1999a) Visualising spatial distributions. In: Longley PA et al. (eds) Geographical Information Systems, principles and technical issues, 2nd ed. John Wiley & Sons, New York, pp 157-173

Kraak MJ (ed) (1999b) Visualisation for exploration of spatial data. International Journal of Geographical Information Systems 13:285-287

Kraak MJ, Edsall R, MacEachren AM (1997a) Cartographic animation and legends for temporal maps: exploration and or interaction. Proceedings of the 7th International Spatial Data Handling Conference, Delft, August 12-16 1996, pp 17-28 [http://www.itc.nl/~kraak/legends/]

Kraak MJ, Edsall R, MacEachren AM (1997b) Kartografische animaties en Legendas voor Temporele Kaarten: Exploratie en/of Enteractie. Kartografisch Tijdschrift 13:13-18

Kraak MJ, MacEachren AM (1994) Visualization of the temporal component of spatial data. In: Waugh TC, Healey RG (eds) Advances in GIS Research. Proceedings, 6th International Symposium on Spatial Data Handling, Edinburgh. Taylor & Francis, London, pp 391-409

Kraak MJ, Müller JC, Ormeling F (1995) GIS-Cartography: visual decision support for spatio-temporal data handling. International Journal of Geographical Information Systems 9:637-645

Kraak MJ, Ormeling FJ (1996) Cartography: visualization of spatial data. Longman, Singapore

Kremers H (1994) Meta-Daten für Geo- und Umweltinformationssysteme. In: Kremers H (ed) Umweltdatenbanken. Metropolis-Verlag, Marburg, pp 203-225

Kriz K (1995) DGM Animation–Eine kartographische Annäherung von 2D bis 4D [www.geog.fu-berlin.de/Karto/texte/kkriz.html]

Kshemkalyani AD (1996) Temporal interactions of intervals in distributed systems. Journal of Computer and System Science 52:287-298

Kumar A, Tsotras VJ, Faloutsos C (1995) Access methods for bi-temporal databases. In: Clifford J, Tuzhilin A (eds) Recent advances in temporal databases. Proceedings of the International Workshop on Temporal Databases, Zürich, Switzerland, 17-18 September 1995 (Workshops in Computing). Springer, Berlin/Heidelberg/New York, pp 235-254

Kunz A (1994) Coordinates for historical maps on the development of German transport since 1835. In: Goerke M (ed) Coordinates for historical maps. A workshop of the Association for History and Computing. European University Institute. Scripta Mercaturae, St. Katharinen, pp 109-120

Kunz A (1999) Verkehrsgeschichte und Computerkartographie. In: Ebeling D (ed) Historisch-thematische Kartographie: Konzepte – Methoden – Anwendungen. Verlag für Regionalgeschichte, Bielefeld, pp 48-59

Kuo TW, Mok AK (1996) Real-time database-similarity semantics and resource scheduling. SIGMOD Record 25:18-22
Kwan MP (1998) Space-time and integral measures of individual accessibility: a comparative analysis using a point-based framework. Geographical Analysis 30:191-216
Lake RW (1993) Planning and applied geography: positivism, ethics and Geographic Information Systems. Progress in Human Geography 17:404-413
Lam KW, Lee VCS, Lam KY, Hung SL (1996) An augmented priority ceiling protocol for hard real-time database systems. Journal of Computing and Information 2:849-866
Lam NSN (1983) Spatial interpolation methods: a review. The American Cartographer 10:129-149
Lane S, Richards K, Chandler J (eds) (1998) Landform monitoring, modelling and analysis. John Wiley & Sons, Chichester
Lang L (1998) Managing natural resources with GIS. ESRI Press, Redlands
Lang L (1999) Transportation GIS. ESRI Press, Redlands
Lang L (2000) GIS for health organizations. ESRI Press, Redlands
Lang R (1984) Probleme bei der zeitlichen und räumlichen Aggregierung topologischer Daten. Geomethodica 9:67-104
Langran G (1988) Temporal GIS design tradeoffs. Proceedings of GIS/LIS '88, San Antonio, pp 890-899
Langran G (1989a) A review of temporal database research and its use in GIS applications. International Journal of Geographical Information Systems 3:215-232
Langran G (1989b) Accessing spatio-temporal data in a temporal GIS. Proceedings of AutoCarto 6, Baltimore. ASPRS/ACSM, Falls Church, pp 191-198
Langran G (1989c) Time in Geographic Information Systems. Unpublished doctoral dissertation. University of Washington, Seattle
Langran G (1990a) Temporal GIS design tradeoffs. Journal of URISA 2:16-25
Langran G (1990b) Tracing temporal information in an automated nautical charting system. Cartography and GIS 17:291-299
Langran G (1991) Issues in implementing a spatio-temporal system. Paper presented at the International Cartographic Conference, Bornemouth
Langran G (1992a) Time in Geographic Information Systems. Technical issues in Geographic Information Systems Series. Taylor & Francis, London
Langran G (1992b) Forest monitoring using temporal GIS techniques. Paper presented at GIS '92, Vancouver, BC
Langran G (1992c) Preparing for a temporal forest GIS. The Compiler 10:18-26
Langran G (1992d) States, events, and evidence: the principal entities of a temporal GIS. Proceedings of GIS/LIS '92, San Jose, pp 417-425
Langran G (1993a) Issues of implementing a spatiotemporal system. International Journal of Geographical Information Systems 7:305-314
Langran G (1993b) Manipulation and analysis of temporal geographic information. Proceedings of Canadian Conference on Geographical Information Systems, Ottawa. The Candian Institute of Geomatics, Ottawa, pp 869-879
Langran G (1995) Updating of real-time spatial systems. Proceedings of 9th Annual Symposium on GIS, Vancouver. GIS World, Fort Collins, pp 354-359
Langran G, Chrisman NR (1988) A framework for temporal geographic information. Cartographica 25:1-14
Larouche C et al. (1997) A practical solution to the geospatial data barrier problem. In: Laurini R et al. (eds) ACM-GIS '97. Proceeding of the Fifth International Workshop on Advances in Geographic Information Systems. ACM, New York, pp 62-66
Laurini R, Thompson D (1992) Fundamentals of spatial information systems (The APIC Series 37). Academic Press, London
Lawrence N (1975) Temporal passage and spatial metaphor. The Study of Time II. Springer, Berlin/Heidelberg/New York
Lee CM, Culhane DP (1998) A perimeter-based clustering index for measuring spatial segregation: a cognitive GIS approach. Environment and Planning B 25:327-343
Lee H, Kemp Z (1998) Supporting complex spatiotemporal analysis in GIS. In: Carver S (ed) Innovations in GIS 5. Selected Papers from the Fifth National Conference on GIS Research UK (GISRUK). Taylor & Francis, London/Bristol, pp 151-161
Lee I (1996) Formal verification, testing and checking of real-time systems. ACM Computing Surveys 28:182

Lee JY, Elmarsi R, Won K (1996) Specification of calendars and time series for temporal databases. Proceedings of the 15th International Conference on Conceptual Modeling, pp 341-356

Lee YC (1995) On data models and schemas. GIS 4:2-6

Lefebre H (1974) The production of space. Basil Blackwell, New York

Leiberich P (1997a) Business Mapping im Marketing. Wichmann, Heidelberg

Leiberich P (1997b) Einführung zum Business Mapping im Marketing. In: Leiberich P (ed) Business Mapping im Marketing. Wichmann, Heidelberg, pp 3-12

Lentz S, Ott T (1996) GIS-gestützte Ermittlung von Wohnbauflächenpotentialen. Standort-Zeitschrift für Angewandte Geographie 20:10-17

Lessing H, Günther O, Swoboda W (1995) Ein objektorientiertes Klassenkonzept für den Umwelt-Datenkatalog (UDK). In: Kremers H, Pillmann W (eds) Raum und Zeit in Umweltinformationssystemen. 9th International Symposium on computer science for environmental protection CSEP '95 (Umwelt-Informatik aktuell). Metropolis-Verlag, Marburg, pp 391-399

Lester M (1990) Tracking the temporal polygon: a conceptual model of multidimensional time for Geographic Information Systems. Paper presented at the Temporal Workshop, University of Maine, 12-13th October 1990, Orono

Leung TYC, Pirahesh H (1995) Querying historical data in IBM DB2 C/S DBMS using recursive SQL. In: Clifford J, Tuzhilin A (eds) Recent advances in temporal databases. Proceedings of the International Workshop on Temporal Databases, Zürich, Switzerland, 17-18 September 1995 (Workshops in Computing). Springer, Berlin/Heidelberg/New York pp 315-331

Leung Y (1997) Intelligent spatial decision support systems (Advances in Spatial Science). Springer, Berlin/Heidelberg/New York

Leykauf J, Albrecht J (1993) Geographical Information Systems and the theory of geographical aspects. In: Czeranka M, Ehlers M (eds): Forschungsberichte zu Geographischen Informationssystemen und Fernerkundung: neue Konzepte und Anwendungen (Vechtaer Studien zur Angewandten Geographie und Regionalwissenschaft 9). Vechtaer Druckerei Verlag, Vechta, pp 49-53

Li JZ, Goralwalla IA, Özsu MT, Szafron D (1997) Modeling video temporal relationships in an object database management system. Proceedings of multimedia computing and modeling. San Jose, pp 80-9

Li KJ (1992) Modélisation d'hypercartes et indexation spatio-temporelle d'objets mobiles. National Institute of applied Science, Ph.D. thesis, Lyon

Li KJ, Badji N, Laurini R (1992a) Modeling moving objects in spatiotemporal databases. Technical Report. Department of Information and Communication Engineering, Seoul

Li KJ, Badji N, Laurini R (1992b) Objets Mobiles: Vers les bases de données spatio-temporelles. Proceedings of the 8th Conference on BDA, Trégastel, France, pp 102-118

Liebig W (1997) Desktop-GIS mit ArcView. Leitfaden für Anwender. Wichmann, Heidelberg

Lin H (1997) Efficient convertion between temporal granularities. Technical Report 19. Time Center, Aalborg

Lin H, Calkins HW (1991) A rationale for spatiotemporal intersection. Proceedings of the ACSM-ASPRS Annual Convention, pp 204-213

Lin KJ, Peng SS (1996) Enhancing external consistency in real-time transactions. SIGMOD Record 25:26-28

Lin L, Risch T (1997) Using a sequential index in terrain-aided navigation. Proceedings of the Sixth International Conference on Information and Knowledge Management (CIKM'97), Las Vegas, November 10-14, 1997

Lin L, Risch T, Sköld M, Badal D (1996) Indexing values of time sequences. Proceedings of the Fifth International Conference on Information and Knowledge Management (CIKM '96), November 12-16, 1996, Rockville,

Linder W (1999) Geo-Informationssysteme. Ein Studien- und Arbeitsbuch. Springer, Berlin/Heidelberg/New York

Litschko T (1999) Raumzeitliche Datenbanken als Basis für GIS-Anwendungen in der Geschichtswissenschaft. In: Ebeling D (ed) Historisch-thematische Kartographie: Konzepte–Methoden–Anwendungen. Verlag für Regionalgeschichte, Bielefeld, pp 167-180

Liu JWS (1996) Validation of timing properties. ACM Computing Surveys 28:184

Lock GR, Stancic Z (ed) (1995) Archaeology and Geographical Information Systems: a European perspective. Taylor & Francis, London

Locke D (1996) On real-time databases. Paper presented at the Fifth International Conference on Information and Knowledge Management, Workshop on databases: active and real-time, Rockyville, MD, November 12-16, 1996

London MG (1997) Time and the mahabharata [http://www.thefirmament.com/Skeets/Research/Papers/mahabharata.html]

Longley P (ed) (1998) Geocomputation: a primer. John Wiley & Sons, Chichester

Longley P, Batty M (eds) (1996) Spatial analysis: modeling in a GIS environment. GeoInformation International, Cambridge

Longley P, Clarke G (ed) (1995) GIS for business and service planning. GeoInformation International, Cambridge

Longley PA et al. (1999) Introduction. In: Longley PA et al. (eds) Geographical Information Systems, principles and technical issues, 2nd ed. John Wiley & Sons, New York, pp 1-16

Longley PA, Goodchild MF, Maguire DJ, Rhind DW (eds) (1999) Geographical Information Systems. Vol 1, principles and technical issues, 2nd ed. John Wiley & Sons, New York/Chichester

Looß R, Schmidt F (1999) Der Einsatz von Geo-Informationssystemen bei einer vergleichenden Analyse von historischen und aktuellen Karten, unpublished paper

Lopez XR (1998) The dissemination of spatial data: a North American-European comparative study on the impact of government information policy (Contemporary Studies). Ablex, Greenwich

Lorentzos N, Johnson R (1988) Requirements specification for a temporal extension to the relational model. Database Engineering 11:204-211

Lorentzos NA (1988) A formal extension of the relational model for the representation and manipulation of generic intervals. Unpublished Ph.D. thesis, Birbeck College, University of London

Lorentzos NA (1993a) Axomatic generalization of the relational model to support valid time data. Internal Report. Informatics Laboratory, Agricultural University of Athens

Lorentzos NA (1993b) DBMS support for time and totally ordered compound data types. Information Systems 17:347-358

Lorentzos NA (1993c) The interval extended relational model and its application to valid time databases. In: Tansel A, Clifford J, Gadis S, Jajodia S, Segev A, Snodgrass R (eds) Temporal databases: theory, design and implementation. Benjamin Cummings, Redwood City, pp 67-91

Lorentzos NA (1994) DBMS support for non-metric measurement systems. IEEE Transactions on Knowledge and Data Engineering 6:945-953

Lorentzos NA, Darwen H (1996) Extension to SQL2 binary operations for temporal data. Proceedings of the 3rd HERMIS Conference, pp 462-469

Lorentzos NA, Johnson RG (1988a) An extension of the relational model to support generic intervals. In: Schmidt JW, Ceri S, Missikoff M (eds) Advances of database technology-EDBT '88. Springer, Berlin/Heidelberg/New York pp 528-542

Lorentzos NA, Johnson RG (1988b) Extending relational algebra to manipulate temporal data. Information Systems 13:289-296

Lorentzos NA, Kollias V (1989) the handling of depth and time intervals in soil-information systems. Computers & Geosciences 15:395-401

Lorentzos NA, Mitsopoulos YG (1996) SQL extension for valid and transaction time support. Technical Report. Informatics Laboratory, Agricultural University of Athens, Athens

Lorentzos NA, Mitsopoulos YG (1997) SQL extension for interval data. IEEE Transactions on Knowledge and Data Engineering 9:480-499

Lowell K (1994) Probabilistic temporal GIS modelling involving more than two map classes. International Journal of Geographical Information Systems 8:73-93

Lowell K, Jaton A (eds) (1999) Spatial accuracy assessment: land information uncertainty in natural resources. Ann Arbor Press, Chelsea

Lum V, Dadam P, Erbe R, Guenaver J, Pistor P, Walch G, Werner H, Woodfill J (1984) Designing DBMS support for the temporal dimension. Proceedings of the SIGMOD '84 Conference, pp 115-130

Luttermann H (1994) Marktstudie und Sachstandsbericht zum Thema: Zeit in geographischen Informationssystemen. Technical Report TR-94001, FAW Ulm [http://www-winfo.uni-siegen.de/winfo/deutsch/publikationen/papers/TGIS-Studie.ps.gz]

Luttermann H, Blobel A (1995) CHRONOS: A spatio-temporal data server for a GIS. In: Kremers H, Pillmann W (eds) Raum und Zeit in Umweltinformationssystemen. 9th International Symposium on Computer Science for Environmental Protection CSEP '95 (Umwelt-Informatik aktuell). Metropolis-Verlag, Marburg, pp 135-142 [http://www-winfo.uni-siegen.de/winfo/deutsch/publikationen/papers/UInf95-09E.ps.gz]

Luttermann H, Grauer M (1999a) Using interactive, temporal visualizations for www-based presentation and exploration of spatio-temporal data. In: Böhlen MH, Jensen CS, Scholl MO (eds) Spatio-temporal database management. Proceedings of the International Workshop STDBM'99 Edinburgh, September 10-11, 1999. Springer, Berlin/Heidelberg/New York, pp 110-118

Luttermann H, Grauer M (1999b) VRML history: storing and browsing temporal 3D-worlds. Proceedings of the VRML 99 Conference, pp. 153-160 [http://www-winfo.uni-siegen.de/vrmlHistory/papers/vrml99pp.pdf.gz]

Lutze G, Schultz A (1995) Daten und Modelle für die Charakterisierung von Agrarlandschaften. In: Kremers H, Pillmann W (eds) Raum und Zeit in Umweltinformationssystemen. 9th International Symposium on Computer Science for Environmental Protection CSEP '95 (Umwelt-Informatik aktuell). Metropolis-Verlag, Marburg, pp 234-241

Lyon JG, McCarthy J (ed) (1995) Wetlands and environmental applications of GIS (Mapping Science). Lewis Publishers, Boca Raton

Lyotrad JF (1984) The postmodern condition: a report on knowledge. Manchester University Press, Manchester

Ma J, Knight B, Peng T (1997) Representing temporal relationships between events and their effects. Fourth International Workshop on Temporal Representation and Reasoning. Proceedings of the Fourth International Workshop on Temporal Representation and Reasoning, TIME '97, May 10-11, 1997, Daytona Beach, pp 148-152

MacDougall EB (1991) Dynamic statistical visualization of Geographic Information Systems. Proceedings of GIS/LIS '91. Atlanta, pp 158-165

MacDougall EB (1992) Exploratory analysis, dynamic static visualization, and Geographic Information Systems. Cartography and GIS 19:237-246

MacEachren A (1993) Time as a Cartographic Variable. In: Unwin D, Hearnshaw H (eds) Visualization in Geographic Information Systems. Belhaven Press, London, pp 115-130

MacEachren AM (1994a) Some truth with maps: a primer on design and symbolization (AAG Resource Paper Series). AAG, Washington, DC

MacEachren AM (1994b) Time as a cartographic variable. In: Hearnshaw HM, Unwin DJ (eds) Visualization in GIS. John Wiley & Sons, Chichester, pp 115-130

MacEachren AM (1995) How maps work: representation, visualization and design. Guilford Press, New York

MacEachren AM (1998a) Cartography, GIS and the World Wide Web. Progress in Human Geography 22:575-585

MacEachren AM (1998b) Visualization–cartography for the 21st century. Proceedigs of the Polish Spatial Information Association Conference, May 1998, Warsaw [http://www.geog.psu.edu/ica/icavis/poland1.html]

MacEachren AM, Boscol FP, Haug D, Pickle LW (1998) Geographic visualization: designing manipulable maps for exploring temporally varying georeferenced statistics. In: Institute of Eletrical and Electronics Engineers (ed) Proceedings of the IEEE Information Visualization Symposium 1998. IEEE, New York, pp 87-94 [http://www.geog.psu.edu/MacEachren/hvisbcd.pdf]

MacEachren AM, Di Biase D (1991) Animated maps of aggregated data: conceptual and practical problems. Cartography and Geographic Information Systems 18:221-229

MacEachren AM, Fraser Taylor DR (1994) Visualization in modern cartography. Elsevier, Oxford

MacEachren AM, Kraak MJ (eds) (1997) Exploratory cartographic visualization. Pergamon Press, Oxford

MacEachren AM, Polsky C, Haug D, Brown D, Boscoe F, Beeddasy J, Pickle L, Marrara M (1997) Visualizing spatial relationships among health, environment, and demographic statistics: interface design issues. Proceedings of the 18th International Cartographic Conference, June 21-27, Stockholm, pp 880-887 [http://www.geo.psu.edu/MacEachren/MacEachrenHTML/NCHS.html]

MacEachren AM, Wachowicz M, Haug D, Edsall R, Masters R (1998) Constructing knowledge from multivariate spatiotemporal data: integrating geographic visualization with knowledge discovery in database methods. International Journal of Geographical Information Science 13:311-334

Mackaness W, Buttenfield B (1991) Incorporating time into geographic process: a framework for analyzing process in GIS. Proceedings of the 15th Conference and 9th General Assembly of the International Cartographic Association, pp 565-574

MacKenzie JW, Hazelton NWJ (1993) Relative scaling of space and time dimensions in Geographic Information Systems. Australian Journal of Geodesy, Photogrammetry and Surveying 59:9-18

MacKenzie JW, Williamson IP, Hazelton B (1995) 4-D adaptive GIS: justification and methodologies. Department of Geomatics, The University of Melbourne [http://www.sli.unimelb.edu.au/research/publications/IPW/4DGISInternal.htm]

MacLennan MJ (1991) The use of a geographic information system for 2nd-order analysis of spatial point patterns. NCGIA Technical Report 91-19

MacMillan B (1989) Modelling through. An afterward. In: MacMillan B (ed) Remodelling geography. Blackwell, Oxford

Madej E (2000) Cartographic design using ArcView GIS. OnWord Press, Albany

Maguire DJ (1989) Computers in geography. Longman, Harlow

Maguire DJ (1994) Spatial data management: multi-user access to continuous spatial databases. AGI '94 Conference Papers, Association for Geographic Information. AGI, London, pp 3.1.1-3.1.6

Maguire DJ, Dangermond J (1991) The functionality of GIS. In: Maguire DJ, Goodchild MF, Rhind DM (eds) Geographical Information Systems: principles and applications. Longman, Harlow, pp 319-335

Maguire DJ, Dangermond J (1994) Future GIS technology. In: Grenn D, Rix D (eds) The AGI sourcebook for geogaphic information systems AGI, London, pp 113-120

Malczewski J (1999) GIS and multicriteria decision analysis. John Wiley & Sons, New York/Weinheim

Mancarella P, Raffaeta A, Turini F (1997) Time in a multi-theory logical framework. Proceedings of the 14th International Workshop on Temporal Representation and Reasoning TIME '97. IEEE, Washington, pp 62-68

Mandelbrot BB (1965) How long is the coast of Britain? Science 156:636-638

Mansfeld W (1998) Satellitenortung und Navigation. Vieweg, Braunschweig/Wiesbaden

Marble D, Peuquet D (1983) Comments on 'automated geography'. The Professional Geographer 35:346-348

Marble DF, DJ Peuquet (eds) (1990) Introductory readings in Geographic Information Systems. Taylor & Francis, London

Margraf O (1991) Quantitative Geographie. Eine Rahmenmethodik zur inhaltlichen Verarbeitung von Attributdateien in GIS. In: Dollinger F, Strobl J (eds) Beiträge zum GIS-Synopsium 3.-5. Juli 1991. Angewandte Geographische Informationstechnologien III (Salzburger Geographische Materialien). Institut für Geographie, Universität Salzburg, pp 117-129

Margraf O (1995) Die landeskundliche Inventarisierung als Grundlage für den Aufbau der Datenbasis für ein länderkundliches GIS (am Beispiel der Buchreihe „Werte der Deutschen Heimat"). In: Dollinger F, Strobl J (eds) Beiträge zum GIS-Synopsium 5.-7. Juli 1995. Beiträge zur Geographischen Informationswissenschaft VII (Salzburger Geographische Materialien). Institut für Geographie, Universität Salzburg, pp 131-136

Mark D (1992) Spatiotemporal relations and GIS: a geographer's view. In: Frank AU, Campari I, Formentini U (eds) Theories and methods of spatio-temporal reasoning in geographic space (LNCS 639). Springer, Berlin/Heidelberg/New York

Mark DM (1999) Spatial representation: a cognitive view. In: Longley PA et al. (eds) Geographical Information Systems, principles and technical issues, 2nd ed. John Wiley & Sons, New York, pp 81-89

Martin C, Sistac J (1996a) An integrity constraint checking method for temporal deductive databases. Paper presented at the Third International Workshop on Temporal Representation and Reasoning TIME '96

Martin C, Sistac J (1996b) Applying transition rules to bitemporal deductive databases for integrity constraint checking. Proceedings of the International Workshop on Logic in Databases (LID '96), pp 117-134

Martin D (1996) Geographic Information Systems. Socioeconomic applications, 2nd ed. Routledge, London/New York

Martin D (1998) Census geography 2001: designed by and for GIS? In: Carver S (ed) Innovations in GIS 5. Selected Papers from the Fifth National Conference on GIS Research UK (GISRUK). Taylor & Francis, London/Bristol, pp 198-209

Martin DJ (1999) Spatial representation: the social scientist's perspective. In: Longley PA et al. (eds) Geographical Information Systems, principles and technical issues, 2nd ed. John Wiley & Sons, New York, pp 71-80

Mason DC, O'Conaill MA, Bell SBM (1994) Handling 4-dimensional geo-referenced data in environmental GIS. International Journal of Geographical Information Systems 8:191-215

Masser I (1998) Governments and Geographic Information. Taylor & Francis, London

Masser I, Blakemore M (1991) Handling geographical information: methodology and potential applications. Longman Scientific & Technical, Harlow

Masuoka P et al. (1996) Techniques for visualizing urban growth using a temporal GIS database. Paper presented at the ASPRS/ACSM Annual Convention and Exhibition, Baltimore, April 22-25, 1996

Matthews JA (1981) Quantitative and statistical approaches to geography. A practical manual. Pergamon Press, Oxford

McCloy KR (1995) Resource management information systems: process and practice. Taylor & Francis, London

McDonald Jampoler S (ed) (1999) GIS in practice. OnWord Press, Santa Fe

McDonald JR, Pullan-Cuthbertson TM (1989) Time and spatial information systems. Proceedings of the Inaugural Colloqium of the Spatial Information Research Centre. University of Otago, pp 79-89

McDonnell R, Kemp K (1996) The international GIS dictionary. John Wiley & Sons, New York

McLaren SB, Braun JK (eds) (1993) GIS applications in mammalogy. Oklahoma Museum of Natural History, Norman

McNaughton GA, McNaughton WP (1999) Deployment of mapping systems in distribution cooperatives. National Rural Electric Cooperative Association, Arlington

McNoleg O (1996) The integration of GIS, remote sensing, expert systems and adaptive cokriging for environmental habitat modeling of the highland haggis using object-oriented, fuzzy-logic and neural-network techniques. Computers & Geosciences 22:585-588

Mecca G, Bonner AJ (1995) Sequences, datalog and transducers. Proceedings of the 14th ACM SIGACT-SIGMOD-SIGART symposium on principles of database systems, pp 23-35

Michailova E (1997) Indices for temporal object databases. Proceedings of the First East-European Symposium on Advance in Database and Information Systems (ADBIS-97), pp 92-94

Michener WK et al. (1994) Environmental information management and analysis: ecosystem to global scales. Taylor & Francis, London

Miller AP (1996) The York archaeological assessment: an investigation of techniques for urban deposit modelling utilising Geographic Information Systems. Unpublished Ph.D. thesis, Department of Archaeology, University of York

Miller EJ (1994) Volumetric visualization of non-point source contamination flow utilizing 4-dimensional geostatistical kriging. Unpublished MSc thesis, Department of Geography, Ohio State University

Miller EJ (1997) Towards a 4D GIS: 4-dimensional interpolation utilizing kriging Innovations in GIS. Taylor & Francis, London, pp 181-197

Miller HJ (1991) Modelling accessibility using space-time prism concepts within Geographical Information Systems. International Journal of Geographical Information Systems 5:287-301

Milne P, Milton P, Smith J (1993) Geographical object-oriented databases: a case study. International Journal of Geographical Information Systems 7:39-55

Minkowski H, Hilbert D (1911) Gesammelte Abhandlungen. Teubner, Leipzig. Reprinted (1967) Chelsea Publications, New York

Mitasova H, Mitas L, Brown WM, Gerdes DP, Kosinovsky I, Baker T (1995) Modeling spatially and temporally distributed phenomena: new methods and tools for GRASS GIS. International Journal of Geographical Information Systems 9:433-446

Mitchell A (1999) The ESRI guide to GIS analysis Volume 1: Geographic patterns & relationships. ESRI Press, Redlands

Mok AK (1996) Firm real-time systems. ACM Computing Surveys 28:185

Molenaar M (1989) Towards a geographic information theory. ITC Journal 1:5-11

Molenaar M (1992) Formal data structures and query spaces. In: Günther O et al. (eds) Konzeption und Einsatz von Umweltinformationssystemen (Informatik Fachberichte 301). Springer, Berlin/Heidelberg/New York, pp 340-362

Molenaar M (1997) Conceptuele ruimtelijke gegevensmodellen. In: Hendriks P, Ottens H (eds) Geografische Informatie Systemen in ruimtelijk onderzoek. Van Gorcum, Assen, pp 21-44

Molenaar M (1998) An introduction to the theory of spatial object modelling for GIS (Research Monographs in GIS). Taylor & Francis, London

Monmonier M (1990) Strategies for the visualization of geographic time-series data. Cartographica 27:30-45

Monmonier M (1991) How to lie with maps. University of Chicago Press, Chicago

Montgomery GE, Schuch HC (1993) GIS Data Conversion Handbook. GIS World Books, Fort Collins

Montgomery LD (1995) Temporal Geographic Information Systems technology and requirements: where we are today. Ohio State University Press, Columbus, Ohio

Morain S (ed) et al. (1996) Raster imagery in Geographic Information Systems. OnWord Press, Santa Fe

Morain SA (ed) (1999) GIS solutions in natural resource management: balancing the technical-political equation. OnWord Press, Santa Fe

Morehouse S (1990) The role of semantics in geographic data modelling. Proceedings of the Fourth International Symposium on Spatial Data Handling, Zurich, Switzerland, July 23-27, pp 689-698

Morrill RL (1977) Efficiency and equity of optimal location models. Geographical Analysis 9:215-226

Morris R, Khatib L (1997) Entities and relations for historical databases. Proceedings of the Fourth International Workshop on Temporal Representation and Reasoning, TIME '97, May 10-11, 1997, Daytona Beach

Mota E, Robertson D, Smaill A (1996) Nature time: temporal granularity in simulation of ecosystems. Journal of Symbolic Comutation 22:665-698

Motakis I, Zaniolo C (1995) Composite temporal events in active databases: a formal semantics. In: Clifford J, Tuzhilin A (eds) Recent advances in temporal databases. Proceedings of the International Workshop on Temporal Databases, Zürich, Switzerland, 17-18 September 1995 (Workshops in Computing). Springer, Berlin/Heidelberg/New York pp 332-351

Motakis I, Zaniolo C (1997) Temporal aggregation in active database rule. In: Peckham, Joan (ed): SIGMOD 1997, Proceedings ACM SIGMOD International Conference on Management of Data, May 13-15, 1997, Tucson

Muehrcke PC (1978) Map use. JP Publications, Madison

Mueller P, Sripada SM (1995) The ChronoBase temporal deductive database system. In: Ong, KL et al. (eds): Knowledge discovery and temporal reasoning in deductive and object-oriented databases (DOOD '95). National University of Singapore, Singapore, pp 83-90

Muggleton S (1991) Inductive logic programming. New Gen Computing 8:295-318

Müller JC, Lagrange JP, Weibel R (eds) (1995) GIS and generalization. Methodology and practice (GISData 1). Taylor & Francis, London/Bristol

Müller M (1992) Entwicklung des Arten-Landschafts-Biotopinformationssystems (ALBIS) als übergreifende Komponente des Umweltinformationssystems Baden-Württemberg. In: Günther O, Schulz KP, Seggelke J (eds) Umweltanwendungen Geographischer Informationssysteme. C.F. Müller, Karlsruhe, pp 64-70

Murray S, Coffman J, Ladd D, Sieg G (1994) Area event systems for forest management in history. Proceedings of the 14th Annual ESRI User Conference, Palm Springs. ESRI Press, Redlands, pp 1359-1375

Mussio L, Forlani G, Crosilla F (eds) (1996) Data acquisition and analysis for multimedia GIS. Springer, Wien

Myrach T (1995) die Schlüsselproblematik bei der Umsetzung temporaler Konzepte in das relationale Datenmodell. Rundbrief des GI-Fachausschusses 5.2, Informationssystem-Archiekturen 2:13-15

Myrach T (1996) Über die Behandlung von Zustands- und Ereignisdaten in Konzepten zeitorientierter Datenhaltung. Rundbrief des GI-Fachausschusses 5.2, Informationssystems-Architekturen 3:41-43

Myrach T (1997a) Realisierung zeitbezogener Datenbanken: Ein Vergleich des herkömmlichen relationalen Datenmodells mit einer temporalen Erweiterung. Wirtschaftsinformatik 39:35-44

Myrach T (1997b) Informationen zu Zeitaspekten betrieblicher Informationssysteme im Internet. Wirtschaftsinformatik 39:181-185

Myrach T (1997c) TSQL2: Der Konsens über eine temporale Datenbanksprache. Informatik-Spektrum 20:143-150

Myrach T, Knolmayer GF, Barnert R (1996) On ensuring keys and referential integrity in the temporal database language TSQL2. Proceedings of the Second international Baltic workshop on databases and information systems 1:171-181

Nagel M, Huber TM, Höring H (1991) Hochinteraktive Graphik und ihre Anwendung auf Daten mit territorialem Bezug. In: Faulbaum F (ed) SoftStat '91 (Advances in Statistical Software 3). Fischer, Jena/New York, pp 177-184

Nascimento M, Dunham MH (1997a) A proposal for indexing bitemporal databases via cooperative b-trees. Proceedings of the International Database Engineering and Applications Symposium, IDEAS 1997, Concordia University, Montreal, August 25-27, 1997, pp 349-357

Nascimento M, Dunham MH (1997b) Indexing valid time database via-tree-the MAP21 approach. Technical Report 97-CSE-08. Southern Methodist University, Dallas

Nascimento MA (1997) A tree-based approach to index transaction time. Technical Report IC-97-09. State University of Campinas, Campinas

Nascimento NA, Dunham MH, Elmasri R (1996) M-IVTT: a practical index for bitemporal databases. Proceedings of the 7th International Conference on Database and eert Systems Applications (DEXA '96) (LNCS 1134). Springer, Berlin/Heidelberg/New York, pp 779-790

Nascimento NA, Dunham MH, Kouramajian V (1996) A multiple tree mapping-based approach for ange indexing. Journal of the Brazilian Computer Society 2:36-46

Newell D (1994) Strategies for managing large AM/FM GIS systems. AGI '94 Conference Papers. AGI, London, pp 3.2.1-3.2.5

Newell R, Easterfield M (1990) Version management-the problem of the long transaction. Technical Paper, Smallworld Systems

Newell RG, Theriault D, Easterfield M (1992) Temporal GIS-modeling the evolution of spatial data in time. Computers & Geosciences 18:427-433

Nievergelt J et al. (eds) (1994) IGIS '94 Geographic Information Systems. International Workshop on Advanced Research in Geographic Information Systems, Monte Verità, Ascona, Switzland. Februar/March 1994. Proceedings (LNCS). Springer, Berlin/Heidelberg/New York

Norvag K, Bratbergsengen K (1997) Log-only temporal object storage. Proceedings of the 8th International Workshop on Database and Expert Systems Applications (DEXA '97), pp 728-733

Nyerges TL (ed) et al. (1995) Cognitive aspects of human-computer interaction for Geographic Information Systems (NATO Asi Series D: Behavioural and Social Sciences 8). Kluwer Academic Publishing, Dordrecht

Obermeyer NJ, Pinto JK (1994) Managing Geographic Information Systems. Guilford Press, New York

O'Conaill MA, Bell SBM, Mason DC (1992) Developing a prototype 4D GIS on a transputer array. ITC Journal 1:47-54

O'Conaill MA, Mason DC, Bell SBM (1993) Spatiotemporal GIS techniques for environmental modelling. In: Mather PM (ed) Geographical information handling-research and applications. John Wiley & Sons, Chichester, pp 103-112

Odland J (1998) Longitudinal analysis of migration and mobility. Spatial behavior in explicitly temporal contexts. In: Egenhofer MJ, Golledge RG (eds) Spatial and Temporal Reasoning in Geographic Informations Systems. Oxford University Press, New York/Oxford, pp 238-259

O'Donoghue PG, Hull MEC (1996) Using timed CSP during object oriented design of real-time systems. Information and Software Technology 38:89-102

Okazaki D (1993) Visualizing geographic data through animation. Proceedings of AutoCarto 11, Minneapolis. ACSM, Bethesda, pp 238-245

O'Kelly MO (1995) Spatial analysis and GIS. In: Fotheringham S, Rogerson P (eds) Spatial Analysis and GIS. Taylor & Francis, London/Bristol, pp 65-79

Olbrich G, Quick M, Schweickert J (1996) Computerkartographie. Eine Einführung in das Desktop Mapping am PC, 2nd ed. Springer, Berlin/Heidelberg/New York

O'Looney J (1997) Beyond maps: GIS decision making in local government (Special Reports). ICMA, Washington, DC

Olsen E, Ramsey R, Winn DS (1994) A modified fractal dimension as a measure of landscape diversity. Photogrammetric Engineering & Remote Sensing 59:1517-1520

O'Neil PE, Ulusoy Ö (1996) Guest editorial: Real-time database systems. Information Systems 21:1-2

Ong KL, Conrad S, Ling TW (1995) Knowledge discovery and temporal reasoning in deductive and object-oriented databases. Paper presented at the DOOD '95 Post-Conference Workshops

Onsrud HJ, Cook DW (ed) (1990) Geographic and Land Information Systems for practicing surveyors: a compendium. ACSM, Bethesda

Onsrud HJ, Hintz RJ (1989) Upgrading boundary information in a GIS using an automated survey measurement management system. Technical Papers, ACSM Annual Convention, pp 275-384

Onsrud HJ, Rushton G (ed) (1995) Sharing geographic information. Center for Urban Policy Research, New Brunswick

Ooi BC (1990) Efficient query processing in Geograhic Information Systems (LNCS 471). Springer, Berlin/Heidelberg/New York

Openshaw S (1979) A geographical solution to scale and aggregation problems in region-building, partitioning and spatial modelling. Transactions of the Institute of British Geographers 2:459-472

Openshaw S (1984) The modifiable areal unit problem. Concepts and techniques in modern geography. Geo-Books, Norwich

Openshaw S (1991a) A spatial analysis research agenda. In: Masser I, Blakemore M (eds) Handling geographical information: methodology and potential applications. Longman, Harlow, pp 18-37

Openshaw S (1991b) A view of the GIS crisis in geography, or using GIS to put Humpy-Dumpty back together again. Environment and Planning A 23:621-828

Openshaw S (1991c) Computer modelling in human geography. In: MacMillan B (ed) Remodelling geography. Blackwell, Oxford, pp 70-88

Openshaw S (1991d) Developing appropriate spatial analysis methods for GIS. In: Maguire DJ, Goodchild MF, Rhind DM (eds) Geographical Information Systems: principles and applications. Longman, Harlow, pp 389-402

Openshaw S (1991e) Handling geographical information. Longman, London

Openshaw S (1992) Further thoughts on geography and GIS: a reply. Environment and Planning A 24:463-466

Openshaw S (1994a) Spatial analysis and GIS. Taylor & Francis, London

Openshaw S (1994b) Two exploratory space-time-attribute pattern analysers relevant to GIS. In: Fotheringham S, Rogerson P (eds) Spatial analysis and GIS. Taylor & Francis, London/Bristol, pp 83-104

Openshaw S, Abrahart RJ (2000) Geocomputation. Taylor & Francis, London

Openshaw S, Taylor JP (1981) The modifiable areal unit problem. In: Wrigley N, Bennett RJ (eds) Quantitative geography: a British view. Routledge & Kegan, London/Boston/Henley, pp 60-69

Openshaw S, Wymer C (1995) Classifying and regionalizing census data. In: Openshaw S (ed) Census user's handbook. GeoInformation International, Cambridge, pp 239-270

Ord JK, Getis A (1995) Local spatial autocorrelation statistics: Distributional Issues and an Application. Geographical Analysis 27:287-306

Orgun M, Faustini AA (1996) The chronolog family of languages. Journal of Symbolic Computation 22:722-724

Orgun MA (1996a) A recursive temporal algebra and temporal completeness. Paper presented at the Third International Workshop on Temporal Representation and Reasoning TIME '96

Orgun MA (1996b) Incorporating an implicit time dimension into he relational model and algebra. Informatique Théorique et Applications/Theoretical informatics and applications 30:231-260

Orgun MA (1996c) On temporal deductive databases. Computional Intelligence 12:235-259

Ormeling F, Van der Schans R (1997) Visualisatie en interactie door kaarten. In: Hendriks P, Ottens H (eds) Geografische informatie systemen in ruimtelijk onderzoek. Van Gorcum, Assen, pp 87-114

Ormsby T (1999) Extending ArcView GIS: with network analyst, spatial analyst and 3D-analyst. ESRI Press, Redlands

Osborne S, Stoogenke M (1989) Integration of a temporal element into a natural resource decision support system. Proceedings of GIS/LIS '89, Washington, DC. ACSM, Bethesda, pp 221-227

Ott T, Swiaczny F (2000) The analysis of cultural landscape change: a GIS approach for handling spatio-temporal data. History and Computing (forthcomming)

Özsoyoglu G, Guruswamy S, Du K, Hou WC (1995) Time-constrained query processing in CASE-DB. IEEE Transactions on Knowledge and Data Engineering 7:865-884

Özsoyoglu G, Snodgrass RT (1995a) Guest editors' introduction to special section on temporal and real-time databases. IEEE Transactions on Knowledge and Data Engineering 7:511-512

Özsoyoglu G, Snodgrass RT (1995b) Temporal and Real-Time Databses: A Survey. IEEE Transactions on Knwoledge and Data Engeneering 7:513-532

Papagno G (1992) Seeing time contribution to the discussion on space-time dynamic. In: Frank AU, Campari I, Formentini U (eds) Theories and methods of spatio-temporal reasoning in geographic space. Springer, Berlin/Heidelberg/New York, pp 29-46

Pápay G (1998) Die Zeitproblematik aus der Sicht raumbezogener historischer Informationssysteme und der Geschichtskartographie. Kartographische Nachrichten 48:177-185

Pápay G (1999) Einsatz von raumbezogenen historischen Informationssystemen in der Landesgeschichtsforschung, erläutert an Beispielen aus der mecklenburgischen und pommerschen Landesgeschichte. In: Ebeling D (ed) Historisch-thematische Kartographie: Konzepte–Methoden–Anwendungen. Verlag für Regionalgeschichte, Bielefeld, pp 24-37

Park J, Lee S (1997) Inheritance of multimedia class having temporal-spatial structures. In: Orlowska, ME, Roberto Z (eds): OOIS'97, 1997 International Conference on Object Oriented Information Systems, 10-12 November 1997, Brisbane, Proceedings. Springer, Berlin/Heidelberg/New York, pp 375-385

Parks D, Thrift N (1980) Times, spaces, and places. John Wiley & Sons, New York

Parthasarathy S (1995) Building temporal constraints into knowledge bases for process control-an examination. Engineering Applications of Artificial Intelligence 8:473

Pascolo P, Brebbia CA (ed) (1998) GIS technologies and their environmental applications. WIT Press, Southampton/Boston

Peckham J, Maryanski F (1988) Semantic data models. ACM Computer Surveys, 20:153-189

Peloux JP, Rigaux p (1995) A loosley coupled interface to an object-oriented geographic database. In: Frank AU, Kuhn W (eds) Spatial information theory-a theoretical basis for GIS. Proceedings COSIT '95 (LNCS 988). Springer, Berlin/Heidelberg/New York, pp 123-137

Perkusich MLB, Perkusich A, Schiel U (1995) Object oriented real-time database design and hierarchical control systems. Proceedings of the First International Workshop on Active and Real-Time Database Systems ARTDB-95 (Workshop on Computing), pp 104-121

Peterson MP (1995) Interactive and animated cartography. Prentice Hall, Englewood Cliffs

Peucker T, Chrisman N (1975) Cartographic data structures. American Cartographer 2:55-69

Peuquet DJ (1984) A conceptual framework and comparison of spatial data models. Cartographica 21:66-113. Reprinted in: Peuquet DJ, Marble DF (eds) (1990) Introductory readings in Geographic Information Systems. Taylor & Francis, London/New York/Philaelphia, pp 250-285

Peuquet DJ (1988a) Representations of geographic space: towards a conceptual synthesis. Annals of the Association of American Geographers 78:375-394

Peuquet DJ (1988b) Toward the definition and use of complex spatial relationships. Proceedings of the Third International Symposium on Spatial Data Handling. IGU, Sydney, pp 211-224

Peuquet DJ (1993) A framework for the representation of spatiotemporal processes in Geographic Information Systems. Paper presented at the International Workshop on an Infrastructure for Temporal Databases, Arlington

Peuquet DJ (1994) It's about time: A conceptual framework for the representation of temporal dynamics in Geographic Information Systems. Annals of the Association of American Geographers 84:441-461

Peuquet DJ (1999) Time in GIS and geographical databases. In: Longley PA et al. (eds) Geographical Information Systems, principles and technical issues, 2nd ed. John Wiley & Sons, New York, pp 91-103

Peuquet DJ (2000) Space-time representation: an overview. In: Heres L (ed): Time in GIS: issues in spatio-temporal modelling (Publications on geodesy 47). Nederlandse Commissie voor Geodesie, Delft, pp 3-12

Peuquet DJ, Duan N (1995) An event-based spatiotemporal data model (EDSTDM) for temporal analysis of geographical data. International Journal of Geographical Information Systems 9:7-24

Peuquet DJ, Marble DF (eds) (1990) Introductory readings in Geographical Information Systems. Taylor & Francis, London

Peuquet DJ, Qian L (1996) An integrated database design for temporal GIS. In: Kraak, MJ (ed): Advances in GIS research II. Proceedings of the Seventh International Symposium on Spatial Data Handling. Taylor & Francis, London, pp 2.1-2.11

Peuquet DJ, Wentz E (1994) An approach for time-based analysis of spatiotemporal data. In: Waugh TC, Healey RG (eds) Advances in GIS research. Proceedings, 6th International Symposium on Spatial Data Handling, Edinburgh. Taylor & Francis, London, pp 489-504

Pickles J (ed) (1995) Ground truth: the social implications of Geographic Information Systems. Guilford Press, New York

Pierau K (1998): Verwaltungsreform und ihre computergerechte Darstellung. Unpublished paper, Zentralarchiv für Empirische Sozialforschung, Köln.

Pigot S (1992) A topological model for a 3D spatial information. Proceedings of the Fifth International Symposium on Spatial Data Handling, Charleston. IGU, Charleston, pp 344-360

Pigot S, Hazelton B (1992) The fundamentals of a topological model for a 4-dimensional GIS. Proceedings of the Fifth International Symposium on Spatial Data Handling, Charleston. IGU, Charleston, pp 580-591

Pissinou N, Snodgrass R, Elmasri R, Mumick I, Oezsu M, Pernici B, Segev A, Theodoulidis B, Dayal U (1994) Towards an infrastructure for temporal databases: report of an invitational ARPA/NSF workshop. SIGMOD Record 23:35-51

Plane DA, Rogerson PA (1994) The geographical analysis of population. With applications to planning and business. John Wiley & Sons, New York/Chichester

Plank T (1994) EQUEL-a database system for the management of environmental data. In: Dollinger F (ed) Angewandte geographische Informationstechnologie 6, Beiträge zum GIS-Symposium 6.-8. Juli 1994 (Salzburger Geographische Materialien 21). Institut für Geographie, Universität Salzburg, pp 551-561

Plewe B (1997) GIS online: information retrieval, mapping and the internet. OnWord Press, Santa Fe

Plöger R (1999) Anwendung Geographischer Informationssysteme (GIS) für historisch-geographische Aufgabenstellungen. In: Ebeling D (ed) Historisch-thematische Kartographie: Konzepte–Methoden–Anwendungen. Verlag für Regionalgeschichte, Bielefeld, pp 9-23

Pollermann K, Rothe H (1998) Interaktive 3D-Grafiken und GIS. In: Strobl J, Dollinger F (eds) Angewandte Geographische Informationsverarbeitung. Beiträge zur AGIT-Symposium, Salzburg 1998. Wichmann, Heidelberg [http://www.geo.sbg.ac.at/agit/papers98/Pollermann/Pollermann.htm]

Preuss H (1995) Informationsvermittlung-eine wichtige Aufgabe moderner Geoinformationssysteme. In: Buziek G et al. (eds) GIS in Forschung und Praxis. Konrad Wittwer, Stuttgart, pp 63-75

Price MF, Heywood DI (ed) (1994) Mountain environments and Geographic Information Systems. Taylor & Francis, London

Price S (1989) Modelling the temporal element in land information systems. International Journal of Geographical Information Systems 3:233-243

Priestnall G, Glover R (1998) A control strategy for automated land use detection: an integration of vector-based GIS, remote sensing and pattern recognition. In: Carver S (ed) Innovations in GIS 5. Selected Papers from the Fifth National Conference on GIS Research UK (GISRUK). Taylor & Francis, London/Bristol, pp 162-175

Proulx MJ, Bédard Y, Létourneau F, Marte C (1996) GEOREP: A WWW customizable georeferenced digital library for spatial data. D-Lib Magazine 2 [http://www.dlib.org/dlib/ december96/canada/12proulx.html]

Qian L, Peuquet D (1988) Evaluating parallel r-tree implementationa on a network of workstations. Proceedings of the 8th International Symposium on Spatial Data Handling, July 11-15, Vancouver, BC, pp 592-602

Qian L, Wachowicz M, Peuquet D, MacEachren A (1997) Delineating operations for visualization and analysis of space-time data in GIS. Proceedings of GIS/LIS '97, Cincinnati, pp 872-877

Qiu K, Hachem NI, Ward MO, Gennert MA (1992) Providing temporal support in data base management systems for global change research. Proceedings of the 6th International Working Conference on Scientific and Statistical Database Management, pp 274-289

Quasten H, Wagner M (1996) Inventarisierung und Bewertung schutzwürdiger Elemente der Kulturlandschaft-eine Modellstudie unter Anwendung eines GIS. Berichte zur deutschen Landeskunde 70:301-326

Quattrochi DA, Goodchild MF (eds) (1997) Scale in remote sensing and GIS. Lewis Publishers, Boca Raton

Quian L (1998) Geographic Foundation Class (GFC) Library [http://www.geog.psu.edu/~qian/ gfc.html]

Raafat H, Yang Z, Gauthier D (1994) Relational spatial topologies for historical geographical information. International Journal of Geographcial Information Systems 8:163-173

Radic U, Jenssen D, McInnes B (1987) On the surging potential of polar ice streams. Antarctic surges-a clear and present danger? Technical Report DOE/ER/60197-H1, US Department of Energy

Rafiei D, Mendelzon AO (1997) Similarity-based queries for time series data. In: Peckham J (ed) Proceedings ACM SIGMOD International Conference on Management of Data, May 13-15, 1997, Tucson, pp 13-25

Ralston BA (1995) Object oriented spatial analysis. In: Fotheringham S, Rogerson P (eds) Spatial analysis and GIS. Taylor & Francis, London, pp 165-185

Ram S (1991) Heterogenous distributed database systems. IEEE Computer 24:7-12

Ram S (1995) Intelligent database design useing the unifying semantic model. Information and Management 19:191-206

Ram S, Park J (1996) Modeling spatial and temporal semantics in a large heterogeneous GIS database environment. Proceedings of the 1996 Americas Conference, Association for Information Systems [http://hsb.baylor.edu/ramsower/ais.ac.96/papers/ram_park.htm]

Ram S, Ramesh V (1995) Schema integration: past, current and future. In: Sheth A (ed) Heterogeneous database systems. Morgan Kaufman, San Mateo

Ramachandran B, MacLeod F, Dowers S (1994) Modelling temporal changes in a GIS using an object-oriented approach. In: Waugh TC, Healey RG (eds) Advanes in GIS Research. Proceedings, 6th International Symposium on Spatial Data Handling, Edinurgh, pp 518-537

Ramamritham K (1996) Predictability: demonstrating real-time properties. ACM Computing Surveys 28:186

Ramamritham K, Sivasankaran RM, Stankoic JA, Towsley DF, Xiong M (1996) Integrating temporal, real-time, and active databases. SIGMOD Record 25:8-12

Ramaswamy S (1997) Efficient indexing for constraint and temporal databases. In: Afrati FN, Kolaitis P (eds) database theory– CDT, '97 6th International Conference, Delphi, January 8-10, 1997, Proceedings. Springer, Berlin/Heidelberg/New York, pp 419-431

Ramsch J, Sosna D (1997) Temporale Datenmodelle und Metainformationen für GIS. Universität Leipzig. Insitut für Informatik Report 10. Selbstverlag, Leipzig

Raper J (1996) Spatio-temporal interpolation in 4-dimensional coastal process models. Proceedings of the Third International Conference/Workshop on Integrating GIS and Environmental Modeling, Santa Fe, January 21-26, 1996 [http://www.sbg.ac.at/geo/idrisi/gis_environmental_modeling/sf_papers/raper_jonathan/raperliv.html]

Raper J (1998) Multi dimensional GIS: extending GIS in space and time. Taylor & Francis, London

Raper J (ed) (1989) Three dimensional applications in Geographical Information Systems. Taylor & Francis, London

Rapper J, Livingstone D (1995) Development of a geomorphological spatial model using object-oriented design. International Journal of Geographical Information Systems 9:359-384

Raper J, Livingstone D (1996) Spatio-temporal interpolation in four dimensional coastal process models. Proceedings of the Third International Conference/Workshop on Integrating GIS and Environmental Modeling, Santa Fe, January 21-26, 1996

Raper JF, Kelk B (1991) 3-dimensional GIS. In: Maguire D, Goodchild M, Rhind D (eds) Geographic Information Systems: principles and applications. Longman, Harlow, pp 299-317

Raper JF, Maguire DJ (1992) Design models and functionality in GIS. Computers & Geosciences 18:387-394

Rase WD (1991) Computer-assisted mapping-the challenges of the nineties. In: Faulbaum F (ed) SoftStat '91 (Advances in Statistical Software 3). Fischer, Jena/New York, pp 473-480

Rase WD (1992) Thematische Kartographie und GIS-Entwicklungen-Anwendungen und Erfahungen in der BfLR. In: Günther O, Schulz KP, Seggelke J (eds) Umweltanwendungen Geographischer Informationssysteme. C.F. Müller, Karlsruhe, pp 211-216

Rase WD (1996) Interpolation von stetigen Oberflächen aus flächenbezogenen Informationen. In: Dollinger F, Strobl J (eds) Angewandte Geographische Informationsverarbeitung VIII. Beiträge zum GIS-Symposium 3.-5. Juli 1996 (Salzburger Geographische Materialien). Institut für Geographie, Universität Salzburg, pp 17-26

Rase WD (1997) Kartographische Anamophosen und verwandte nichtlineare Abbildungen für thematische Karten. In: Grünreich D (ed) GIS und Kartographie im multimedialen Umfeld. Grundlagen, Anwendungen und Entwicklungstrends (Kartographische Schriften). Kirschbaum, Bonn, pp 83-89

Razavi AH (1997) Arcview GIS/Avenue developer's guide. OnWord Press, Santa Fe
Razavi AH, Warwick V, Alexander J (2000) ArcView GIS/Avenue Programmer's Reference, 3rd ed. OnWord Press, Albany
Rennison E, Strausfeld L (1995) The millennium project: constructing a dynamic 3-D virtual environment for exploring geographically, temporally and categorically organized historical information. In: Frank AU, Kuhn W (eds) Spatial information theory-a theoretical basis for GIS. Proceedings COSIT '95 (LNCS 988). Springer, Berlin/Heidelberg/New York, pp 69-91
Reutter E, Stefan B, Peithmann O (1999) GIS-Analyse des historischen Bodennutzungswandels. In: Asch K (ed) GIS in Geowissenschaften und Umwelt. Springer, Berlin/Heidelberg/New York, pp 143-162
Rex B, Risch J (1996) Animation query language for the visualization of temporal data. Proceedings of the Third International Conference/Workshop on Integrating GIS and Environmental Modeling, Santa Fe, January 21-26, 1996 [http://www.sbg.ac.at/geo/idrisi/gis_environmental_modeling/sf_papers/rex_bruce/rex_bruce.html]
Rhind D (1989) Computing, geography and the outside world. In: MacMillan B (ed) Remodelling geography. Blackwell, Oxford
Rhind D, Taylor D (1989) Cartography past, present and future. ICA, London
Ridd MR (1991) Spatial and temporal scale issues related to the integrating of GIS and remote sensing. Proceedings of the ACSM-ASPRS Annual Convention on Remote Sensing and Geographic Information Systems, pp 167-173
Riekert WF (1992) Fernerkundung als wissensbasierte Verarbeitung geographischer Information. In: Günther O, Riekert WF (eds) Wissensbasierte Methoden zur Fernerkundung der Umwelt. Wichmann, Karlsruhe, pp 9-36
Rietsema van Eck JR (1993) Analyse van transportnetwerken in GIS voor sociaal-geografisch onderzoek (Nederlandes Geografische Studies). KNAG, Utrecht
Ritsema I (2000) Time in relation to geoscientific data. In: Heres L (ed): Time in GIS: issues in spatio-temporal modelling (Publications on geodesy 47). Nederlandse Commissie voor Geodesie, Delft, pp 57-69
Rikkers R, Molenarr M, Stuiver J (1994) A query orientated implementation of a topologic data structure for 3-dimensional vector maps. International Journal of Geographical Information Systems 8:243-260
Ripple WJ (ed) (1989) Fundamentals of Geographic Information Systems: a compendium. American Society for Photogrammetry and Remote Sensing, Bethesda
Ris F (1991) Toward a scientific GIS. Proceedings of GIS/LIS '91, Atlanta. ACSM, Bethesda, pp 740-751
Robertson GG, McKinlay JD, Card SK (1991) Cone Trees: animated 3D visualizations of hierarchical information. In: Robertson et al. (eds) Proceedings of the CHI '91 Conference, pp 189-194
Robey D, Newman M (1996) Sequential patterns in information systems development: an application of a social process model. ACM Transactions on information systems 14:30-63
Robinson AH (1956) The necessity of weighting values in correlation analysis of area data. Annals of the Association of American Geographers 46:233-236
Robinson GM (1998) Methods and techniques in human geography. John Wiley & Sons, Chichester
Robinson VB, Tom H (1994) Towards Sql database language extensions for Geographic Information Systems. Silicon Press, Summit
Robinson WS (1950) Ecological correlation and the behaviour of individuals. American Sociological Review 15:351-357
Roche C (1996) How to structure data: the object oriented approach (a survey). In: Dubois JE, Gershon N (eds) Modelling complex data for creating information. Springer, Berlin/Heidelberg/New York, pp 3-8
Roman GC (1990) Formal specification of geographic data processing requirements. IEEE Transactions on Knowledge and Data Engineering 2:370-380
Ross J (1985) Detecting land use change on Omaha's urban fringe using a Geographic Information System. Proceedings of AutoCarto 7, Washington, DC. ASPRS/ACSM, Falls Church, pp 463-471
Rossignol J, Wandsnider L (eds) (1992) Space, time and archaeological landscapes. Plenum Press, New York
Ruf T (1997) Scientific & statistical databases-Datenbankeintsatz in der multidimensionalen Datenanalyse. Vieweg, Braunschweig/Wiesbaden

Rumbaugh J (1993) Objektorientiertes Modellieren und Entwerfen. Hanser, München

Ryan NS (1998) Field Note desktop. An experimental spatio-temporal information system. Paper presented at the 4th International Colloquium on Computing and Archaeology, Bilbao.

Sack RD (1980) Conceptions of space in social thought. A geographical perspective. Macmillan, London

Salzberg B (1994) On indexing spatial and temporal data. Paper MA 02115, Northeastern University, Boston

Salzberg B, Tsotras VJ (1997) A comparison of access methods for temporal data. Technical Report 18. Time Center, Aalborg

Samet H (1993) Application of spatial data structures. Computer Graphics, image processing, and GIS. Addison Wesley, Reading

Samet H (1995) Spatial data structures. In: Kim W (ed) Modern database systems: the object model, interoperability and beyond. ACM, New York, pp 361-385

Sample VA (ed) (1994) Remote sensing and GIS in ecosystem management: American forests. Island Press, Washington, DC

Sandhu J (1990) Combining exploratory data analysis and scientific visualization in the study of very large space-time data. Department of Geography, Ohio State University, Columbus

Sarda NL (1997) Modeling valid time: an efficient representation. In: Topor RW, Tanaka K (eds) Proceedings of the Fifth International Conference on Database Systems for Advanced Applications (DASFAA), Melbourne, April 1-4, 1997, pp 363-370

Sarda NL, Reddy PCSP (1997) Handling of alternatives and events in temporal databases. Technical Report. IIT, Bombay

Saurer H, Behr FJ (1997) Geographische Informationssysteme. Eine Einführung. Wissenschaftliche Buchgesellschaft, Darmstadt

Saurer H, Dikau R (1999) GIS and earth surface systems. Analysis and modelling of the natural environment. Bornträger, Berlin/Stuttgart

Savitsky BG, Thomas E, Lacher J (eds) (1998) GIS methodologies for developing conservation strategies: tropical forest recovery and wildlife management in Costa Rica (Biology and Resource Management). Columbia University Press, New York

Sayer A (1989) The new regional geography and problems of narrative. Environment and Planning D 7:253-276

Schellerer J (1992) Objektorientierung in Geographischen Informationssystemen anhand aktueller Anwendungsbeispiele. GIS 4:14-19

Schilling P (1991) Variabler Thesaurus-eine Schlüsselfunktion für die zukünftige Informationsverarbeitung in einer Verwaltung. In: Günther O, Kuhn H et al. (eds) Workshop Umweltinformatik 1990. Tagungsband Teil II. Forschungsinstitut für anwendungsorientierte Wissensverarbeitung an der Universität Ulm FAW-B-91002 / II. FAW, Ulm, pp 11-17

Schleifer P, Sun Y, Patel D (1997) Temporal business objects: A waste of time? In: Orlowska ME, Zicari R (eds) OOIS'97, International Conference on Object Oriented Information Systems, 10-12 November 1997, Brisbane, Proceedings. Springer, Berlin/Heidelberg/New York, pp 247-260

Schleuß U, Heinrich U, Scholle D, Zölitz-Möller R (1999) Erstellung und Auswertung einheitlicher gemeinsamer Geometrien im Rahmen der Ökosystemforschung. Geographische Informationssysteme 12:10-16

Schmidt D, Kotz Dittrich A, Dreyer W, Marti R (1995) Time Series, a neglected issue in temporal database research? In: Clifford J, Tuzhilin A (eds) Recent advances in temporal databases. Proceedings of the International Workshop on Temporal Databases, Zürich, Switzerland, 17-18 September 1995 (Workshops in Computing). Springer, Berlin/Heidelberg/New York, pp 214-232

Schmitt D (1996) Kopplung des Raster-GIS IDRISI mit einem Fuzzy-Klassifikationsmodell zur Modellierung eines unscharfen räumlichen Entscheidungsprozesses. In: Lorup E, Strobl J (eds) IDRISI '96 (Salzburger Geographische Materialien 25). Institut für Geographie, Universität Salzburg, CD-ROM

Schneider M (1995) Spatial data types for database systems. Unpublished thesis. Fernuniversität Hagen, Hagen

Schneider M (1997) Spatial data types for database systems: finite resolution geometry for Geographic Information Systems (LNCS 1288). Springer, Berlin/Heidelberg/New York

Schneider R, Kriegel HP (1992) Indexing the spatiotemporal monitoring of a polygonal object. Proceedings of the Fifth International Symposium on Spatial Data Handling, Charleston, pp 200-209

Schott D (1999) HIST–Entwicklung und Anwendung eines historischen Geo-Informationssystems für die Stadtforschung. In: Ebeling D (ed) Historisch-thematische Kartographie: Konzepte–Methoden–Anwendungen. Verlag für Regionalgeschichte, Bielefeld, pp 38-47

Schumann H, Müller W (2000) Visualisierung. Grundlagen und allgemeine Methoden. Springer, Berlin/Heidelberg/New York

Schumm S, Lichty R (1965) Time, space, and causality in geomorphology. American Journal of Science 263:110-119

Schwalb E, Dechter R (1996) Processing disjunctions of temporal constraints. Paper presented at the Third International Workshop on Temporal Representation and Reasoning TIME '96

Schwartz S, Tochtermann K, Riekert WF, Hermsdörfer D, Hansen J (1999) Open geospatial warehouse-Raumbezogene Dienstleistungen im Intra-/Internet der Stadt Köln. In: Strobl J, Blaschke T (eds) Angewandte Geographische Informationsverarbeitung, Beiträge zum AGIT-Symposium, Salzburg 1999. Wichmann, Heidelberg [http://www.geo.sbg.ac.at/agit/papers99/Schwartz/schwartz.htm]

Schwarz von Raumer HG (1997) GIS-gestützte mehrkriterielle Bewertungsverfahren in der Stadtentwicklungsplanung. GIS 6:13-19

Scott JA (1993) The potential for using a Geographic Information System to model the spread of bracken (steridium aquilinum). Unpublished MSc thesis, Department of Biology, University of York

Segev A, Shoshani A (1993) A temporal data model based on time sequences. In: Tansel et al. (eds) Temporal databases: theory, design and implementation. Benjamin Cummings, Reading, pp 248-270

Sen A, Smith TE (1995) Gravity models of spatial interaction behavior. Springer, Berlin/Heidelberg/New York

Seshadri P, Livny M, Ramakrishnan R (1996) The design and implementation of a sequence database systems. Proceedings of the 22nd International Conference on Very Large Databases, pp 99-110

Sha L (1996) Real-time in the real world. ACM Computing Surveys 28:187

Shahar Y (1997) Knowledge-based temporal interpolation. Proceedings of the 4th International Workshop on Temporal Representation and Reasoning, TIME '97, May 10-11, 1997, Daytona Beach, pp 102-111

Shekhar S et al. (1997) Experiences with data models in Geographic Information Systems. Communications of the ACM 40:154-168

Shen Q (1998) Location characteristics of inner-city neighborhoods and employment accessibility of low-wage workers. Environment and Planning B 25:345-365

Shepherd IDH (1995) Innovations in GIS II. Taylor & Francis, London

Shibasaki R (1994) Handling spatio-temporal uncertainties of geo-objects for real-time update of GIS databases from multi-source data. In: Molenaar M, De Hoop S (eds) Advanced geographic data modelling 1994 (AGDM '94), Spatial data modelling and query languages for 2D and 3D applications, Proceedings, Delft, 12-14 September (Publications on geodesy 40), pp 228-242

Shimakawa H (1996) Active and real-time databases for industrial plants. Paper presented at the Fifth International Conference on Information and Knowledge Management, Workshop on databases: active and real-time, Rockyville, MD, November 12-16, 1996

Shoham Y, Goyal N (1988) Temporal reasoning in artificial intelligence. In: Shrobe HE (ed) Exploring artificial intelligence. Kaufmann, San Mateo, pp 419-438

Sieber R, Bär HR(1997) Atlas der Schweiz-Multimediale Version. Adaptierte GIS-Techniken und qualitative Bildschirmgraphik. In: Grünreich D (ed) GIS und Kartographie im multimedialen Umfeld. Grundlagen, Anwendungen und Entwicklungstrends (Kartographische Schriften). Kirschbaum, Bonn, pp 67-77

Siebert E (1996) Raumordnungs-, Kommunal- und Infrastrukturplanung. Lektion 10: Deponiestandortsuche. In: Bumann E, Bachhuber R, Schaller J (eds) ArcView. GIS-Arbeitsbuch. Wichmann, Heidelberg, pp 175-188

Singer C (1991) SICAD geographic database management-concepts and facilities. In: Schilcher M (ed) Geo-Informatik. Anwendungen, Erfahrungen, Tendenzen. Siemens Nixdorf, Berlin/München, pp 71-83

Singh VP, Fiorentino M (ed) (1996) Geographical Information Systems in hydrology (Water Science and Technology Library 26). Kluwer, Dordrecht

Singhroy VH et al. (eds) (1996) Remote sensing and GIS for site characterization: applications and standards (ASTM Special Technical Publication 1279). ASTM, West Conshohocken

Sinton D (1978) The inherent structure of information as a constraint to analysis: mapped thematic data as a case study. In: Dutton G (ed) Harvard Papers on GIS. Addison-Wesley, Reading

Sivasankaran R, Ramamritham K, Stankovic JA, Townsley DF (1995) Data placement, logging and recovery in real-time active databases. Proceedings of the First International Workshop on Active and Real-time Database Systems ARTDB-95 (Workshop on Computing), pp 226-241

Sivasankaran RM, Stankovic JA, Townsley DF, Purimetla B, Ramamritham K (1996) Priority assignment in real-time active databases. The VLDB Journal 5:19-34

Sivertun A (1993) Geographical Information Systems (GIS) as a tool for analysis and communication of multidimensional data. Geographical Reports 10. Department of Geography, University of Umea, Umea

Six HW, Widmayer P (1988) Spatial searching in geometric databases. Proceedings of the International Conference on Data Engineering, pp 496-503

Skjellaug B (1997a) Temporal data: time and object databases. Research report 245. University of Oslo, Department of Informatics, Oslo

Skjellaug B (1997b) Temporal data: time and relational databases. Research report 246. University of Oslo, Department of Informatics, Oslo

Skjellaug B, Berre AJ (1997a) A uniform temporal geographic data model framework. SINTEF report STF40 A97046. SINTEF Telecom and Informatics, Oslo

Skjellaug B, Berre AJ (1997b) Multi-dimensional time support for spatial data models. Technical Report, Research report 253, May. University of Oslo, Department of Informatics, Oslo

Skowron A, Suraj Z (1996) A parallel algorithm for real-time decision making: a rough set approach. Journal of Intelligent Information Systems 7:5-28

Slocum TA (1999) Thematic cartography and visualization. Prentice-Hall, Upper Saddle River

Smith N (1989) Geography as museum: private history and conservative idealism. In: Entrikin J, Brunn S (eds) Reflections on R. Hartshorne's 'The nature of geography'. AAG, Washington DC, pp 91-120

Smith N (1992) History and philosophy of geography: real wars, theory wars. Progress in Human Geography 16:257-271

Smith T (1992) Towards a logic-based language for modeling and database support for spatio-temporal domains. Proceedings of the Fifth International Conference on Spatial Data Handling, pp 592-601

Smith TR (1994) On the integration of database systems and computational support for high-level modelling of spatio-temporal phenomena. In: Worboys MF (ed) Innovations in GIS. Taylor & Francis, London, pp 11-24

Smith TR, Frank A (1990) Report on workshop on very large spatial databases. Journal of Visual Languages and Computing 1:291-309

Smith TR, Su J, Agrawal D, El Abbadi A (1993) Database and modelling systems for the earth sciences. IEEE Transactions on Knowledge and Data Engineering 16:33-37

Smyth CS (1998) A representational framework for geographic modeling. In: Egenhofer MJ, Golledge RG (eds) Spatial and temporal reasoning in Geographic Informations Systems. Oxford University Press, New York/Oxford, pp 191-213

Smyth S (1992) A representational framework for route planning in space and time. Proceedings of the Fifth International Symposium on Spatial Data Handling, pp 692-701

Snodgrass R (1987) The temporal query language tquel. ACM Transactions on Database Systems 12:247-298

Snodgrass R, Ahn I (1985) A taxonomy of time in databases. Proceedings of the SIGMOD '85 Conference. ACM, New York, pp 236-245

Snodgrass R, Ahn I (1986) Temporal databases. IEEE Computer 19:35-42

Snodgrass R, Böhlen MH, Jensen CS, Steiner A (1996) Adding transaction time to sql/temporal. Change proposal ISO/IEC JTC1/SC21/WG3 DBL MAD-147r2, ANSI X3H2-96-502r1.

Snodgrass RT (1992) Temporal databases. In: Frank AU, Campari I, Formentini U (eds) Theories and methods of spatio-temporal reasoning in geographic space. Springer, Berlin/Heidelberg/New York, pp 22-57

Snodgrass RT (1999) Developing time-oriented database applications in SQL. Morgan Kaufmann, San Francisco

Snodgrass RT, Ahn I, Ariav G, Batory D, Clifford J, Dyreson CE, Elmasri R, Grandi F, Jensen CS (1994) Tsql2 language specification [ftp://ftp.cs.arizona.edu/tsql/tsql2/spec.pdf]

Snodgrass RT, Böhlen M, Jensen CS, Steiner A (1997) Transitioning temporal support in TSQL2 to SQL3 Technical Report 8. Time Center, Aalborg

Snodgrass RT, Clifford J, Dayal U, Segev A (1995) Panel: Whither TSQL3? In: Clifford J, Tuzhilin A (eds) Recent advances in temporal databases. Proceedings of the International Workshop on Temporal Databases, Zürich, Switzerland, 17-18 September 1995 (Workshops in Computing). Springer, Berlin/Heidelberg/New York, p 355

Son S (1996) Issues and non-issues in real-time database systems. Workshop on databases: active and real-time. Paper presented at the Fifth International Conference on Information and Knowledge Management, Workshop on databases: active and real-time, Rockyville, MD, November 12-16, 1996

Son SH, David R, Thuraisingham BM (1996) Improving timeliness in realtime secure database systems. SIGMOD Record 25:29-33

Song X, Liu JWS (1995) Maintaining temporal consistency: pessimistic vs. optimistic concurrency control. IEEE Transactions on Knowledge and Data Engineering 7:786-796

Southall H, Gregory I (2000) The Great Britain Historical GIS. Paper presented at the workshop „Mapping Europe's Historic Boundaries and Borders“, European University Institute, Florence/Italy, June 2-4, 2000 [http://www.geog.port.ac.uk/hist-bound/papers/gbhgis_paper.htm]

Southall H, White B (1997) Creating an electronic historical atlas of Britain. Geocal 16:3-6

Sowa JF (1984) Conceptual structures: information processing in mind and machine. Addison-Wesley, Reading

Spery L, Claramunt C, Libourel T (1999) A lineage metadata model for the temporal management of a cadastre application. In: Tjoa AM, Cammelli A, Wagner RR (eds) Proceedings of the International workshop on Spatio-temporal Models and Language SDTSML '99. The IEEE Computer Society, Florence, pp 466-474

Srikant R, Agrawal R (1996) Mining sequential patterns: generalizations and performance improvements. Advances in Database Technology-EDBT'96, Fifth International Conference on Extending Database Technology, Avignon, March 25-29, 1996, Proceedings (LNCS 1057). Springer, Berlin/Heidelberg/New York, pp 3-17

Sripada S (1992) On snapshot, rollback, historical and temporal deductive databases. Proceedings of Computer and Data Management, Bangalore, pp 83-93

Sripada SM (1993) A metalogic programming approach to reasoning about time in knowledge bases. In: Bajcsy R (ed) Proceedings of the 13th International Joint Conference on Artificial Intelligence, Chambéry, August 28 – September 3, 1993, pp 860-865

Sripada SM (1995) Efficient implementation of the event calculus for temporal database applications. In: Sterling E (ed) Logic programming, Proceedings of the Twelfth International Conference on Logic Programming, June 13-16, 1995, Tokyo. MIT Press, pp 99-113

Sripada SM, Mueller P (1995) The generalized chronobase temporal data model meta-logics and logic Programming. Proceedings of the 12th International Conference on logic programming (ICLP 1995) June . National University of Singapore, Singapore, pp 83-90

Stachowiak H (1965) Gedanken zu einer allgemeinen Theorie der Modelle. Studium Generale 18:432-463

Stafford DB (ed) (1991) Civil engineering applications of remote sensing and Geographic Information Systems: Proceedings: Sheraton City Center Washington, DC, May 14th-16th. American Society of Civil Engineers, New York

Stanek H, Smith N, Giordano A (1995) Modellierung und Normierung von Datenqualität im GIS. In: Dollinger F, Strobl J (eds) Beiträge zum GIS-Symposium 5.-7. Juli 1995. Beiträge zur Geographischen Informationswissenschaft VII (Salzburger Geographische Materialien). Institut für Geographie, Universität Salzburg, pp 265-274

Stankovic JA (1996) Strategic directions in real-time and embeded systems. ACM Computing Surveys 28:751-763

Star J, Estes JE (1990) Geographic Information Systems: an introduction. Prentice Hall, Englewood Cliffs

Stead SD (1998) Temporal dynamics and Geographic Information Systems. In: Egenhofer MJ, Golledge RG (eds) Spatial and temporal reasoning in Geographic Informations Systems. Oxford University Press, New York/Oxford, pp 214-219

Stein A, Frits W.T. Penning De Vries (eds) (1999) Data models in action: methodological issues in production ecology (Current Issues in Production Ecology). Kluwer Academic Publishers, Dordrecht/London

Steinborn W (1998) Wirtschaftsmacht der Koordinaten-warum nutzen immer mehr Firmen GIS? Geographische Informationssysteme 11:3-4

Steiner A, Norrie MC (1997a) A temporal extension to a generic object data model. Technical Report 15. Time Center, Aalborg

Steiner A, Norrie MC (1997b) Implementing temporal database in object-oriented systems. Proceedings of Fifth International Conference on Database Systems for Advanced Applications (DASFAA '97), pp 381-390

Steiner A, Norrie MC (1997c) Temporal object role modelling. Proceedings of 9th International Conference on Advanced Information Systems Engineering (CAiSE '97), pp 245-258

Stewart CAJ, McLeod F, Dowers S (1994) Towards the implementation of a spatio-historical database: the Tcobject conceptual meta model. Proceedings of the GIS Research UK 1994 Conference, Leicester, pp 254-262

Steyaert LT, Goodchild MF (1994) Integrating Geographic Information Systems and environmental simulation models: a status review. In: Michener WK, Brunt JW, Stafford SG (eds) Environmental information management and analysis: ecosystem to global scales. Taylor & Francis, London, pp 333-356

Stillwell JCH (ed) et al. (1999) Geographical information and planning (Advances in Spatial Science). Springer, Berlin/Heidelberg/New York

Stock O (ed) (1997) Spatial and temporal reasoning. Kluwer Academic Publishers, Dordrecht

Storch H (1998) Die Bedeutung von Metadaten zur Steuerung von WebGIS-Anwendungen im Kontext von Umweltinformationssystemen (UIS). In: Strobl J, Dollinger F (eds) Angewandte Geographische Informationsverarbeitung. Beiträge zum AGIT-Symposium Salzburg '98. Wichmann, Heidelberg, pp 374-379

Story PA, Worboys MF (1995a) A design support environment for spatio-temporal database applications. In: Frank AU, Kuhn W (eds) Spatial information theory-a theoretical basis for GIS. Proceedings COSIT '95 (LNCS 988). Springer, Berlin/Heidelberg/New York, pp 413-430

Story PA, Worboys MF (1995b) An object-oriented model of time. Technical Report. Department of Computer Sciene, Keele University, Keele

Stoyenko AD, Marlowe TJ (1995) A case for better language and compiler support for real-time database systems. Proceedings of the First International Workshop on Active Real-Time Database Systems ARTDB-95 (Workshops in Computing), pp 46-49

Strobl J (1995) Grundzüge der Metadatenorganisation für GIS. In: Dollinger F, Strobl J (eds) Beiträge zum GIS-Symposium 5.-7. Juli 1995. Beiträge zur Geographischen Informationswissenschaft VII (Salzburger Geographische Materialien). Institut für Geographie, Universität Salzburg, pp 275-286

Studer R (1986) Modeling time aspects of information systems. Proceedings of the International Conference on Data Engineering. IEEE, New York, pp 364-372

Su H (1997) Semantics-based time-alignment operations in temproal query processing and optimization. Information Science 103:37-70

Swiaczny F, Ott T (1999) Das Problem der Modellierung historischer Prozesse mit Geographischen Informationssystemen. Historical Social Research 24:75-100

Sykes JM (1996a) More elements of type period. Technical Report ISO/IEC JTC 1/SC 21/WG 3 DBL MCI-004

Sykes JM (1996b) Periods of integers. Technical Report ISO/IEC JTC 1/SC 21/WG 3 DBL MAD-151

Tagashira T, Amagasa T, Aritsugi M, Kanamori Y (1997) Interval-based representations of spatio-temporal concepts. The 9th Conference on Advanced Information Systems Engineering CAiSE '97 (LNCS 1250), pp 231-244

Taina J, Raatikainen K (1996) Rodain: a real-time object-oriented database system for telecom. Paper presented at the Fifth International Conference on Information and Knowledge Management, Workshop on Databases: Active and Real-Time, Rockyville, November 12-16, 1996

Tang Q (1992) From description to analysis: an electronic atlas for spatial data exploration. Proceedings of ASPRS/ACSM/RT 1992, Washington, DC. ACSM, Bethesda, pp 455-463

Tansel A, Clifford J,Gadia S, Jajodia S, Segev A, Snodgrass RE (1993) Temporal databases: theory, design and implementation (Database Systems and Applications Series). Benjamin Cummings, Redwood City

Tansel AU (1997) Temporal relational data model. IEEE Transactions on Knowledge and Data Engineering 9:464-479

Tansel AU, Tin E (1997) The expressive power of temporal relational query languages. IEEE Transactions on Knowledge and Data Engineering 9:120-134

Taylor DRF (ed) (1991) Geographic Information Systems: the microcomputer and modern cartography (Modern Cartography 1). Pergamon, Oxford

Taylor DRF (ed) (1998) Policy issues in modern cartography (Modern Cartography 3). Pergamon, New York

Taylor DRF (1999) Future directions for multimedia cartography. In: Cartwright et al. (eds) Multimedia cartography. Springer, Berlin/Heidelberg/New York, pp 315-326

Taylor PJ, Overton M (1991) Further thoughts on geography and GIS. Environment and Engineering A 23:1087-1094

Terenziani P (1996) Toward an ontology dealing with periodic events. Proceedings of the 12th European conference on artificial intelligence (ECAI '96), pp 43-47

Terenziani P (1997a) Integrating calendar-dates and qualitative temporal constraints in the treatment of periodic events. IEEE Transactions on Knowledge and Data Engineering 9:763-783

Terenziani P (1997b) Qualitative and quantitative temporal constraits about numerically quantified periodic events. Proceedings of Fourth International Workshop on Temporal Representation and Reasoning TIME '97. IEEE, Washington, pp 94-101

Teunissen PJG, Kleusberg A (eds) (1998) GPS for Geodesy. Springer, Berlin/Heidelberg/New York

Theodoridis Y, Papdias D (1995) Reange queries involving spatial relations: a performance analysis. In: Frank AU, Kuhn W (eds) Spatial information theory-a theoretical basis for GIS. Proceedings COSIT '95 (LNCS 988). Springer, Berlin/Heidelberg/New York, pp 537-551

Theodoridis Y, Vazirgannis M, Sellis T (1996) Spatio-temporal indexing for large multimedia applications. In: Institute of Eletrical and Electronics Engineers (ed) Proceedings of the IEEE Multimedia Systems Conference. IEEE, New York, pp 441-448

Theodoulidis CI, Loucopoulos P (1991) The time dimension in conceptual modelling. Information Systems 16:273-300

Thériault M, Claramunt C (1999) La modélisation du temps et des processus dans les SIG: Un moyen d'intégration pour la recherche interdisciplinaire. Revue Internationale de Geomatique 9:67-103

Thill JC (ed) (1999) Spatial multicriteria decision making and analysis: a Geographic Information Sciences Approach. Ashgate, Aldershot

Thomas EN, Anderson DL (1965) Additional comments on weightin values in correlation analysis of areal data. Annals of the Association of American Geographers 55:492-505

Thomas S, Seshadri S, Haritsa JR (1996) Integrating standard transactions in firm real-time database systems. Information Systems 21:3-28

Thrift N (1977) An introduction to time geography (Concepts in modern geography). Geo-Abstracts, London

Timm C et al. (1998) Möglichkeiten und Grenzen von GIS-Komponententechnologie in der Geodatenpoduktion. In: Strobl J, Dollinger F (ed) Angewandte Geographische Informationsverarbeitung. Beiträge zum AGIT-Symposium Salzburg '98. Wichmann, Heidelberg, pp 380-386

Timpf S, Frank A (1997) Metadaten-mulitmedial? In: Dollinger F, Strobl J (eds) Angewandte Geographische Informationsverarbeitung IX. Beiträge zum AGIT-Symposium 2.-4. Juli 1997 (Salzburger Geographische Materialien). Institut für Geographie, Universität Salzburg, pp 163-172

Tobler W (1979) Cellular geography. In: Gaile S, Olsson G (eds) Philosophy in geography. Reidel, Dordrecht, pp 379-386

Tobler W (1989) Accuracy of spatial databases. Taylor & Francis, London

Tobler WR (1970) A computer movie: Simulation of population change in the Detroit Region. Economic Geography 46:234-240

Tom CH (1997) Spatio-temporal GIS data sources. Paper presented at the 9th Annual GIS in the Rockies Conference, Denver

Toman D (1997) Point-based temporal extension of SQL. In: Bry F, Ramakrishnan R, Ramamohanaroa K (eds) Deductive and object-oriented databases, Fifth International Conference, DOOD'97, Montreux, Switzerland, December 8-12, 1997, Proceedings (LNCS 1341). Springer, Berlin/Heidelberg/New York, pp 103-121

Toman D, Chomicki J, Rogers DS (1994) Datalog with integer periodicity constrains. Proceedings International Logic Programming Symposium, pp 189-203

Toman D, Niwinski D (1996) First-order queries over temporal databases inexpressible in temporal logic. In: Apers PMG, Bouzeghoub M, Gardarin G (eds) Advances in database technology-EDBT'96, Fifth International Conference on Extending Database Technology, Avignon, March 25-29, 1996, Proceedings (LNCS 1057). Springer, Berlin/Heidelberg/New York, pp 307-324

Tomlinson RF (1987) Review of North American experience of current and potential uses of GIS. Handling geographic information. HMSO, London, pp 153-161

Tomlinson RF (1990) Geographic Information Systems-a new frontier. In: Peuquet DJ, Marble DF (eds) Introductory readings in Geographic Information Systems. Taylor & Francis, London, pp 18-29

Toppen F, Ottens H (1997) Gegevens voor een GIS. In: Hendriks P, Ottens H (eds) Geografische informatie systemen in ruimtelijk onderzoek. Van Gorcum, Assen, pp 65-86

Torp K, Jensen CS, Böhlen M (1997) Layered temporal DBMSs-concepts and techniques. In: Topor RW, Tanaka K (eds) Proceedings of the Fifth International Conference on Database Systems for Advanced Applications (DASFAA), Melbourne, April 1-4, 1997, pp 371-380

Torp K, Jensen CS, Snodgrass RT (1997) Stratum approaches to temporal DBMS implementation Technical Report 5. Time Center, Aalborg

Torp K, Mark L, Jensen CS (1998) Efficient differential timeslice computation. IEEE Transactions on Knowledge and Data Engineering 10:599-611

Torp K, Snodgrass RT, Jensen CS (1997) Correct and efficient timestamping of temporal data Technical Report 4. Time Center, Aalborg

Tryfona N, Hadzilacos T (1997) Evaluation of database modeling methods for Geographic Information Systems. CHOROCHRONOS-a network for spatiotemporal database systems, Technical Report CH-97-05 [http://www.dbnet.ntua.gr/~choros/TRs/1997/5/report.ps.gz]

Tryfona N, Jensen CS (1998) Conceptual data modeling for spatiotemporal applications. Chorochronos, Technical Report CH-98-08. National Technical University of Athens

Tsotras VJ, Jensen CS, Snodgrass RT (1997) A notation for spatiotemporal queries. Technical Report 10. Time Center, Aalborg

Tuckey JW (1994) Exploratory data analysis. Addison Wesley, Reading

Tufte ER (1990) Envisioning information. Graphics Press, Cheshire

Turner MG, Dale VH, Gardner RH (1989) Predicting across scales: theory development and testing. Landscape Ecology 3:245-252

Tuzhilin A, Padmanabhan B (1996) Pattern discovery in temporal databases: a temporal logic approach. Second International Conference on Knowledge Discovery and Data Mining (KDD-96), pp 351-354

UK Experts (1995) IXSQL: an alternative approach on temporal data. Technical Report ISO/IEC JTC 1/SC 21/WG 3 DBL YOW-031

Ulusoy Ö (1995a) A study of two transaction-processing architectures for distributed real-time data base systems. Journal of Systems and Software 31:97-10

Ulusoy Ö (1995b) An annotated bibliography on real-time database systems. ACM SIGMOD Record 24:40-45

Ulusoy Ö (1995c) Research issues in real-time database systems: survey paper. Information Science 87:123-151

Ulusoy Ö (1997) An evaluation of network access protocols for distributed real-time database systems. Journal of Systems and Software 37:49-60

Ulusoy Ö, Buchmann AP (1996) Exploiting main memory DBMS features to improve real-time concurrency control protocols. SIGMOD Record 25:23-25

Unwin A (1996) Exploring spatio-temporal data. In: Fischer ME, Scholten HJ, Unwin D (eds) Spatial analytical perspectives on GIS (GISData 4). Taylor & Francis, London/Bristol, pp 101-110

Unwin A, Hofmann H (1998) New interactive graphics tools for exploratory analysis of spatial data. In: Carver S (ed) Innovations in GIS 5. Selected Papers from the Fifth National Conference on GIS Research UK (GISRUK). Taylor & Francis, London/Bristol, pp 46-55

Unwin DJ (1994) Cartography, ViSC and GIS. Progress in Human Geography 18:516-522

Unwin DJ (1995) Geographical Information Systems and the problem of 'error and uncertainty'. Progress in Human Geography 19:549-558

Unwin DJ (1996) GIS, spatial analysis and spatial statistics. Progress in Human Geography 20:540-551

Unwin DJ, Hearnshaw HM (1994) Visualization in Geographical Information Systems. John Wiley & Sons, Chichester

Upton G, Fingleton B (1985) Spatial data analysis by example. Point pattern and quantitative data. John Wiley & Sons, Chichester
Valsecchi P, Claramunt C, Peytchev E (1999) OSIRIS: an inter-operable system for the integration of real time traffic data within GIS. Computer Environment and Urban Systems Journal 23:245-257
Valsecchi P, Claramunt C, Spaccapietra S (1999) A cooperative real-time GIS for traffic management. In: Zhang Y, Rusinkiewicz M, Kambayashi Y (eds) Cooperative database systems and applications '99, Proceedings of the Second Interntaional Symposium on Cooperative Database Systems for Advanced Applications (CODAS '99), Wollongong, Australia, March 27-28, 1999. Springer, Singapore, pp 303-315
Van Benthem J (1996) Remodeling temporal geometry. Paper presented at the Third International Workshop on Temporal Representation and Reasoning TIME '96
Van der Bercken J, Seeger B, Widmayer P (1997) A generic approach to bulk loading multidimensional index structures. Proceedings of the 23rd International Conference on Very Large Databases VLDB'96, pp 406-415
Van der Kooy JW (1994) Object-orientation and GIS users. In: EGIS-Foundation (ed) Fifth European Conference and Exhibition on GIS. EGIS, Paris, pp 1214-1223
Van Kreveld M et al. (eds) (1997) Algorithmic foundations of Geographic Information Systems (LNCS 1340). Springer, Berlin/Heidelberg
Van Lammeren R, Roos-Klein Lankhorst J (1997) GIS-toepassing in de landschapsecologie. In: Hendriks P, Ottens H (eds) Geografische informatie systemen in ruimtelijk onderzoek. Van Gorcum, Assen, pp 183-194
Van Nouhuys J, Glücklich J, Bethge A (1998) Bestandsaufnahme und Analyse von Umweltanwendungen Geographischer Informationsysteme (GIS) in Bund und Ländern. Umweltbundesamt Texte. Umweltbundesamt, Berlin
Van Oosterom P (1999) Spatial access methods. In: Longley PA et al. (eds) Geographical Information Systems, principles and technical issues, 2nd ed. John Wiley & Sons, New York, pp 385-400
Van Oosterom P (2000) Time in cadastral maps. In: Heres L (ed): Time in GIS: issues in spatio-temporal modelling (Publications on geodesy 47). Nederlandse Commissie voor Geodesie, Delft, pp 36-45
Van Teefelen P, Kwant T (1999) Tussen belofte en werkelijkheid: GIS-toepassing in ontwikkelingslanden. Geografie 8:36-39
Van Wijngaarden F et al. (1997) Map integration-update propagation in a multi-source environment. In: Laurini R et al. (eds) ACM-GIS '97. Proceeding of the Fifth International Workshop on Advances in Geographic Information Systems. ACM, New York, pp 71-76
Vance D et al. (1998) Inside Autodesk world. OnWord Press, Santa Fe
Vanhaute E (1994) The quantitative database of Belgian municipalitis (19th/20th centuries). From diachronic worksheets to historical maps. In: Goerke M (ed) Coordinates for historical maps. A workshop of the Association for History and Computing. Scripta Mercaturae, St. Katharinen, pp 162-175
Varma H, Boudreau H, Prime W (1990) A data structure for spatio-temporal databases. International Hydrographic Review 67:71-92
Varman PJ, Verma RM (1997) An efficient multiversion access structure. IEEE Transactions on Knowledge and Data Engineering 9:391-409
Vasiliev I (1990) Examples of the treatment of time as a variable on maps. Temporal GIS workshop. University of Maine, Orono
Vasiliev IR (1997) Mapping time. Cartographica 49:1-51
Vassilakis C, Lorentzos NA, Georgiadis P (1995) Transactions support in a temporal DBMS. In: Clifford J, Tuzhilin A (eds) Recent advances in temporal databases. Proceedings of the International Workshop on Temporal Databases, Zürich, Switzerland, 17-18 September 1995 (Workshops in Computing). Springer, Berlin/Heidelberg/New York pp 255-271
Vassilakis C, Lorentzos NA, Georgiadis P (1996) Implementation of transaction and concurrent users support in a temporal DBMS. Technical Report. Agricultural University of Athens, Informatics Laboratory, Athens
Vassilakis C, Lorentzos NA, Georgiadis P, Mitsopoulos YG (1996) ORES: design and implementation of a temporal DBMS. Technical Report. Agricultural University of Athens, Informatics Laboratory, Athens
Vckovski A (1998) Interoperable and distributed processing in GIS (Research Monographs in GIS). Taylor & Francis, London/Bristol

Velho L (1990) Adaptive polygonization of implicit surfaces. Proceedings of Ausgraph '90, pp 339-343

Verbyla DL, Chang, KT (1997) Processing digital images in Geographic Information Systems: a tutorial featuring Arcview and Arc/Info. OnWord Press, Santa Fe

Vervloet JAJ (1994) Zum Stand der angewandten historischen Geographie in den Niederlanden. Berichte zur deutschen Landeskunde 68:445-458

Vieille L et al. (1992) The EKS-V1 System. In: Vornokov A (ed) Logic programming and automated reasoning, International Conference LPAR '92, St. Petersburg, July 15-20, 1992, Proceedings (LNCS 624). Springer, Berlin/Heidelberg/New York, pp 504-506

Vieux BE (1991) Geographic Information Systems and non-point source water quality and quantity modelling. Hydrological Processes 5:101-113

Visvalingam M (1989) Cartography, GIS and maps in perspective. Cartographic Journal 26:26-32

Voigtmann A (1997) An object-oriented database kernel for spatio-temporal geo-applications. Unpublished thesis, Institut für Informatik. Westfälische Wilhelms-Universität, Münster

Vrana R (1989) Historical data as an explicit component of land information systems. International Journal of Geographical Information Systems 3:33-49. Reprinted in: Peuquet DJ, Marble DF (eds) (1990) Introductory readings in Geographic Information Systems. Taylor & Francis, London/New York/Philadelphia, pp 286-302

Vrbsky SV (1996) A data model for approximate query processing of real-time databases. Data & Knowledge Engineering 21:79-102

Wachowicz M (1999) Object-oriented design for temporal GIS. Taylor & Francis, London

Wachowicz M (2000) The role of geographic visualisation and knowledge discovery in spatio-temporal data. In: Heres L (ed): Time in GIS: issues in spatio-temporal modelling (Publications on geodesy 47). Nederlandse Commissie voor Geodesie, Delft, pp 13-26

Wachowicz M, Healey RG (1994) Towards temporality in GIS. In: Worboys MF (ed) Innovations in GIS. Taylor & Francis, London, pp 105-115

Wachowicz M, MacEachren AM (1998) Spatio-temporal data visualization support for knowledge construction. Paper presented at the Annual Meeting of the Association of American Geographers, March 26-29, Fort Worth

Wadsworth R, Treweek J (1999) GIS for ecology: an introduction. Addison Wesley Longman, Harlow

Walford N (1999) Making more of maps: geography and Geographical Information Systems. Geography 363:129-138

Wang F (1995) Timing behaviour analysis for real-time systems. In: Institute of Eletrical and Electronics Engineers (ed) Proceedings of the 10th Annual IEEE Symposium on logic in Computer Science. IEEE, New York, pp 112-122

Wang F, Hall BG (1996) Fuzzy representation of geographical boundaries in GIS. International Journal of Geographical Information Systems 10:573-590

Wang XS, Bettini C, Brodsky A, Jajodia S (1997) Logical design for temporal databases with multiplegranularities. ACM Transactions on Database Systems 22:115-170

Weber CR (1998) The representation of spatio-temporal variation in GIS and cartographic displays: The case for sonification and auditory data representation. In: Egenhofer MJ, Golledge RG (eds) Spatial and temporal reasoning in Geographic Informations Systems. Oxford University Press, New York/Oxford, pp 74-84

Weibel R, Buttenfield BP (1992) Improvement of GIS graphics for analysis and decision-making. International Journal of Geographical Information Systems 6:223-245

Weismiller RA, Kristof SJ, Scholz DK, Anuta PF, Momin SA (1977) Change detection in coastal zone environments. Photogrametric Engineering and Remote Sensing 43:1533-1539

Wessels C, Van Oosterom P, Molenaar M (1997) Het structureren van ruimtelijke informatie. In: Hendriks P, Ottens H (eds) Geografische informatie systemen in ruimtelijk onderzoek. Van Gorcum, Assen, pp 45-64

West R, Schwan K, Tacic I, Ahamad M (1997) Exploiting temporal and spatial constraints on distributed shared objects. Paper presented at the 17th International Conference on Distributed Computer Systems, 28-30 May 1997, Baltimore

Westcott K, Brandon RJ (2000) Practical applications of GIS for archaelogists: a predictive modeling kit (GISData). Taylor & Francis, London/New York

Whigham PA (1993) Hierarchies of space and time. In: Frank AU, Campari I (ed) Spatial information theory. A theoretical basis for GIS. Proceedings COSIT '93 (LNCS 7169). Springer, Berlin/Heidelberg/New York, pp 190-201

White B (1997) Electronic atlases: in theory and in practice. Bulletin of the Society of Cartographers 31:5-10

Whitener A, Davis J (1998) Minding your business with MapInfo. OnWord Press, Santa Fe

Whitener A, Ryker B (1997) Mapbasic developer's guide. OnWord Press, Santa Fe

Wickenkamp V, Beins-Franke A, Duttmann R (1996) Ansätze zur GIS-gestützten Modellierung dynamischer Systeme und Simulation ökologischer Prozesse. In: Dollinger F, Strobl J (eds) Angewandte Geographische Informationsverarbeitung VIII (Salzburger Geographische Materialien). Institut für Geographie, Universität Salzburg, pp 51-60

Wijsen J (1995) Design of temporal relational databases based on dynamic and temporal functional dependencies. In: Clifford J, Tuzhilin A (eds) Recent advances in temporal databases. Proceedings of the International Workshop on Temporal Databases, Zürich, Switzerland, 17-18 September 1995 (Workshops in Computing). Springer, Berlin/Heidelberg/New York pp 61-76

Wijsen J, Meersman R (1997) On the complexity of mining temporal trends. Paper presented at SIGMOD '97 workshop on research issues on data mining and knowledge discovery (DMKD '97)

Wilke T (1995) Qualitätsaspekte bei der Nutzung von Geo-Informationssystemen. In: Buziek G et al. (eds) GIS in Forschung und Praxis. Konrad Wittwer, Stuttgart, pp 141-154

Wilkinson GG et al (1986) Review of integrated geo-information systems techniques-final report on ESA contract NO 6084/84/D/IS (SC)

Williams GJ, Woods SG (1993) Representing expectations in spatial information systems. In: Abel D, Ooi BC (eds) Advances in spatial databases. Third International Synopsium, SSD '93, Proceedings (LNCS 692). Springer, Berlin/Heidelberg/New York, pp 465-476

Wilmersdorf E (1996) Die Abbildung der Dynamik in einem kommunalen Informationsystem. In: Maier F, Kriz K (eds) Kartographie im multimedialen Umfeld 5. Univeristät Wien, Institut für Geographie, Wien, pp 135-146

Winfree AT (1980) The geometry of biological time. Springer, Berlin/Heidelberg/New York

Winnige N (1999) Alphabetisierung in der Frühen Neuzeit, oder: Wie visualisiere ich raumbezogene historische Daten? In: Ebeling D (ed) Historisch-thematische Kartographie: Konzepte – Methoden – Anwendungen. Verlag für Regionalgeschichte, Bielefeld, pp 82-97

Wirth E (1979) Theoretische Geographie. Teubner, Stuttgart

Wolf A (1989) The DASDBA Geo-Kernel, concepts, experiences, and the second step, design and implementation of large spatial databases. Proceedings of the First Symposium on Spatial Databases, Berlin, pp 67-88

Won J, Elmasri R, Kwak JD (1997) Comparing the expressiveness of temporal data models based on time dimension representation. Paper presented at the Third International Workshop on Next Generation Information Technologies and Systems (NGITS '97), Neve Ilan, Israel

Wong JSK, Mitra S(1996) A nonblocking timed atomic commit protocol for distributed realtime database systems. Journal of Systems and Software 34:161-170

Wood D, Fells J (1986) Design on signs: myth and meaning in maps. Cartographica 23:54-103

Wood D, Fells J (1993) The power of maps. Routledge, London

Wood M, Brodlie K (1994) ViSC and GIS: some fundamental considerations. In: Hearnshaw HM, Unwin DJ (ed) Visualization in GIS. John Wiley & Sons, Chichester, pp 3-8

Woods RL (1976) Aspects of the scale problem in the calculation of segregation indices: London and Birmingham, 1961 and 1971. Tijdschrift voor Economische en Socicale Geografie 67:169-174

Woodsford P (1995) The significance of object-orientation for GIS [http://www.laser-scan.com/papers/ooforgis.htm]

Worboys MF (1990a) Reasoning about GIS using temporal and dynamic logics. NCGIA Technical Report 90.4, Santa Barbara

Worboys MF (1990b) Temporal logic. Temporal relations workshop. University of Maine, Orono

Worboys MF (1990c) The role of modal logics in the description of knowledge in a Geographical Information System. Cognitive and linguistic aspects of geograhic space. NATO ASI

Worboys MF (1992a) A model for spatio-temporal information. Proceedings of the Fifth International Symposium on Spatial Data Handling, pp 602-611

Worboys MF (1992b) Object-oriented models of spatiotemporal information. Proceedings of GIS/LIS '92, San Jose, pp 2:825-834

Worboys MF (1993) Putting time into GIS. AGI'93 Conference Papers. Association for Geographic Information, London, pp 3.8.1-3.8.5

Worboys MF (1994a) GIS: a computing perspective. Taylor & Francis, London

Worboys MF (1994b) A unified model for spatial and temporal information. The Computer Journal 37:26-34

Worboys MF (1994c) Unifying the spatial and temporal components of geographical information. In: Waugh TC, Healey RG (eds) Advances in GIS Research. Proceedings, 6th International Symposium on Spatial Data Handling, Edinburgh. Taylor & Francis, London, pp 505-517

Worboys MF (1998a) A generic model for spatio-bitemporal geographic information. In: Egenhofer MJ, Golledge RG (eds) Spatial and temporal reasoning in Geographic Information Systems. Oxford University Press, New York/Oxford, pp 25-39

Worboys MF (1998b) Modelling changes and events in dynamic spatial systems with references to socio-economic units. Paper presented at the ESF GISData Conference on Modelling Change in Socio-Economic Units.

Worboys MF (1999) Relational databases and beyond. In: Longley PA (ed) Geographical Information Systems, principles and technical issues, 2nd ed. John Wiley & Sons, New York, pp 373-384

Worboys MF, Hearnshaw HM, Maguire DJ (1990) Object-oriented data modelling for spatial databases. International Journal of Geographical Information Systems 4:369-383

Wright DJ, Bartlett DJ (eds) (2000) Marine and coastal Geographical Information Systems (Research Monographs in Geographic Information Systems). Taylor & Francis, London

Wright JK (1936) A method of mapping densities of population. With cape cod as an example. The Geographical Review 26:103-110

Wu F, Webster CJ (1998) Simulation of land development through the integration of cellular automata and multicriteria evaluation. Environment and Planning B 25:103-126

Wuu G, Dayal U (1992) A uniform model for temporal object-oriented databases. Proceedings of the International Conference on Data Engineering, Tempe, pp 584-593

Xiao Q, Raafar H, Gauthier D (1989) A temporal/spatial database structure for remotely sensed image data management within GIS. Proceedings of GIS/LIS '89 Washington, DC. ACSM, Bethesda, pp 116-122

Xiong M, Stankovich JA, Ramamritham K, Towsley DF, Sivasankaran R (1996) Maintaining temporal consistency: issues and algorithms. In: Lin KJ (ed) Proceedings of the First International Workshop on Real-Time Databases: Issues and Applications, March 7-8, 1996, Newport Beach, pp 1-6

Xu X, Han J, Lu W (1990) RT-Tree: an improved R-Tree index structure for spatiotemporal databases. Proceedings of the Fourth International Symposium on Spatial Data Handling, pp 1040-1049

Yang J, Widom J (1997) Maintaining temporal views over non-historical information sources for data warehousing. Technical Report. Stanford Database Groups, Stanford

Yau M, Srihari SN (1983) A hierarchical data structure for multi dimensional images. Communications of the ACM 26:504-515

Yearsley CM, Worboys M (1994) A deductive model of complexareal objects. Proceedings of the GIS Research UK 1994 Conference, Leicester, pp 80-86

Yearsley CM, Worboys MF (1995) Innovations in GIS. Taylor & Francis, London

Yeh T (1992) Temporal aspects of geographical databases. Proceedings of the Third EGIS, Munich, March 23-26 1992, pp 320-328

Yuan M (1994a) Representation of wildfire in Geographic Information Systems. Unpublished Ph.D. thesis, Department of Geography, State University of New York, Buffalo

Yuan M (1994b) Wildfire conceptual modeling for building GIS space-time models. Proceedings of GIS/LIS '94, Phoenix, pp 860-869

Yuan M (1996) Temporal GIS and spatio-temporal modeling. Proceedings of the Third International Conference/Workshop on Integrating GIS and Environmental Modeling, Santa Fe, January 21-26, 1996 [http://www.ncgia.ucsb.edu/conf/SANTA_FE_CD-ROM/sf_papers/ yuan_may/may.html]

Yuan M, Albrecht J (1995) Structural analysis of geographic information and GIS operations from a user's perspective. In: Frank AU, Kuhn W (eds) Spatial information theory-a theoretical basis for GIS. Proceedings COSIT '95 (LNCS 988). Springer, Berlin/Heidelberg/New York, pp 107-122

Zagel B (2000) GIS in Verkehr und Transport. Wichmann, Heidelberg

Zeiler M (1997) Inside Arc/INFO V8, 2nd ed. OnWord Press, Santa Fe

Zeiler M (1999) Modeling our world: The ESRI guide to Geodatabase Design. ESRI Press, Redlands

Zhao F (1997a) Temporal GIS and its application to transportation. Transportation Research Board, Washington, DC

Zhao F (1997b) Transportation applications of temporal GIS. Proceedings of the 1997 ESRI user conference [http://www.esri.com/library/userconf/proc97/proc97/to450/pap427/p427.htm]

Zhao Y (1997) Vehicle location and navigation systems (Artech House ITS Series). Artech House, Boston/London

Zimány E, Parent C, Spaccapietra S, Pirotte A (1997) TERC+: a temporal conceptual model. Paper presented at the International Symposium on Digital Media Information Base (DMIB '97), Nara, Japan, November 1997

Zipf A (1996) Einführung in GIS und ARC/INFO (Heidelberger Geographische Bausteine 13). Universität Heidelberg, Geographisches Institut, Heidelberg

Zölitz-Möller R, Keßler M (1996) Erprobung von ATKIS-Geometrien als Grundlage für konsistente Geofachdaten des Umweltressorts in Schleswig-Holstein. Kartographische Nachrichten 46:167-175

Zurek T (1997a) Optimal interval partitioning for temporal databases. Proceedings of the Third Basque International Workshop on Information Technology (BIWIT '97), pp 140-147

Zurek T (1997b) Optimisation of partitioned temporal joins. Proceedings of the 15th British National Conference on Databases (BNCOD '97). Springer, Berlin/Heidelberg/New York, pp 101-115

Zurek T (1997c) Parallel processing of temporal joins. Informatica 22:153-166

Zurek T (1997d) Parallel temporal joins. In: Dittrich KR (ed) Datenbanksysteme in Büro, Technik und Wissenschaft. Proceedings. Springer, Berlin/Heidelberg/New York, pp 269-278

9.2
Internet links

The following tables present a survey of useful internet sites that deal with theories and applications of temporal GIS research. In this case of course comprehensiveness can neither be intended nor accomplished. It is rather a question of providing entry points and favourable examples which are supposed to demonstrate the affluence of information available on the Internet. However, the enormous dynamics of the Internet generates not only new sources frequently. Interesting pages may not longer be kept up-to-date and addresses may disappear overnight. The contents itself is often still available under a new address. Hopefully you will find a notice at the old address, which is, however, not always guaranteed. Therefore it is possible, that you end up in Nirvana while trying to open a document mentioned below. On the other hand you will be compensated by further references on the working pages.

There is no need to type in all the web addresses manually. Simply visit our website at http//www.time-gis.de where you will find an updated version of this link list for your convenience.

9.2.1
Research projects

Title	Tempest
URL	http://www.geog.psu.edu/tempest/
People	Donna Peuquet et al.
Institution(s)	Penn State University, Department of Geography
Description	Tempest is an experimental GIS that focuses on the management and analysis of large volumes of spatio-temporal data.

Title	The Apola Project
URL	http://www.geovista.psu.edu/grants/apoala/index.htm
People	Donna Peuquet et al.
Institution(s)	Penn State University, Geographic Visualization Science, Technology (GEOVISTA)
Description	The Apoala Project has been undertaken to develop an integrated spatio-temporal geographic information, visualization, and analysis system with the goal of supporting complex exploratory analysis of environmental data.

Title	Spatiotemporal GIS: An investigation into a Distributed Multimedia Implementation
URL	http://www.cs.ru.ac.za/research/mikep/
People	Peter Clayton, George Wells, Michael Preston

Institution(s) Rhodes University, Department of Computer Science
Description This project aims to explore the distributed multimedia side to GIS applications, in particular Virtual GIS, using a standard WEB Browser as the user interface. The limitations of available bandwidth and processing power for delivering "real-time" Virtual GIS, using 3-D GIS data sets and distributed multimedia streams as input, will be investigated. Temporal GIS is currently under investigation as an extension to this project.

Title Great Britain Historical GIS Project
URL http://www.geog.port.ac.uk/gbhgis/
People Humphrey Southall, Ian Gregory
Institution(s) University of Portsmouth, Department of Geography
Description This project is creating a historical GIS for Great Britain covering the period from the late 1830s, when modern statistical data collection can be said to have started, until the early 1970s, when data starts to become available in digital form. The GIS has two major components; accurate boundary data for the changing administrative areas of Britain linked to a major database of social, economic, and electoral statistics from throughout the period. This, we hope, will become a major resource for researchers interested in any or all of the periods, as well as providing the backbone for the production of a major new atlas covering British history over the past 200 years.

Title Temporal Urban Mapping
URL http://edcwww2.cr.usgs.gov/umap/umap.html
Institution(s) U.S. Geological Survey, EROS Data Center
Description Temporal urban mapping develops a historical perspective of urban development by combining a variety of data sources into an integrated, multi-scale, and multi-resolution database. The database provides the baseline information to model and predict regional patterns of urbanization. Temporal urban mapping relies on modern mapping techniques, such as remote sensing and geographic information systems (GIS), to capture information from both historical and modern records. The database highlights the profound changes to the landscape that have incrementally developed over time.

Title Spatio-Temporal GIS Analysis for Environmental Health
URL http://www.spatial.maine.edu/~max/nih.html
People David M. Mark, Ling Bian and Peter Rogerson, John Vena, Max Egenhofer
Institution(s) University at Buffalo, University of Maine, National Institute of Environmental Health Sciences

Description This research project focuses on the extraction of health-related information from geo-spatial lifelines, which capture individuals' locations in geographic space at regular or irregular temporal intervals. The objectives of the work are to develop and test the theory of geo-spatial lifelines in the environmental health sciences.

Title Busstop – a spatio-temporal information system
URL http://www.gfy.ku.dk/~tk/busstop/index.htm
People Thomas Knudsen
Institution(s) National Survey and Cadastre, Denmark
Description Busstop, an acronym for Basic Utilities for Spatial and Spatio-Temporal Operations, is a spatio-temporal information system (STIS), intended as an aid in the process of integrating heterogeneous spatio-temporal datasets, primarily remote-sensing data. By focusing equally on the spatial and temporal aspects of geodata, Busstop should become a more powerful tool than traditional geographical information systems (GIS) for studies of dynamic systems (such as the ocean and the atmosphere).

Title TimeMap Project
URL http://www.archaeology.usyd.edu.au/research/time_map/
Institution(s) University of Sydney, Archaeological Computing Laboratory
Description The TimeMap Project tackles the isolation of cultural data through an explicit methodological approach to recording cultural data in time and space. By defining a conceptual mapping (materialised as a metadata table) between an explicit spatio-temporal data model (the Snapshot-Transition model) and the data actually recorded by any particular project (in whatever format), our methodology allows existing resources to be integrated into a unified structure and interrogated together. The methodological approach is complemented by the development of pragmatic data storage, indexing, access and display methods. As a result we now have a working prototype of a federated system consisting of a time-aware mapping front-end, automated central catalogue and a distributed heterogeneous population of datasets.

Title Ein historisch-geographisches Informationssystem zur Geschichte Preußens
URL http://www2.hu-berlin.de/fgp/hgis.htm#Grundlagen
People Norbert Winnige
Institution(s) Forschungsinstitut für die Geschichte Preußens e.V.
Description This concept of a historical GIS of Prussia follows the example of the Great Britain Historical GIS, described above. (in German)

Title An Integrated Approach for Effective Representation and Analysis of Space/Time Environmental Data
URL http://es.epa.gov/ncerqa_abstracts/grants/96/high/peuque.html
People Donna J. Peuquet, Anand Sivasubramaniam, Brent Yarnall, Alan MacEachren
Institution(s) U.S. Environmental Protection Agency, National Center for Environmental Research
Description The goal of this research project is to develop the fundamental capabilities needed as the basis for an integrated temporal GIS / visualization environment that facilitates science and decision support for assessment of environmental problems and potential solutions. Toward this overall goal, the objective of this research is to design and implement a prototype Temporal GIS with integrated multivariate spatio-temporal capabilities. The prototype will include: (1a) a fully-integrated database model which incorporates object-based, location-based and time-based representations of environmental data in a tightly-coupled manner, (1b) the use of parallelism for highly efficient storage, retrieval and manipulation of data, (2) a flexible database query protocol linked to a graphical user interface (GUI) which facilitates multi-dimensional and multi-scale query and analysis, (3) a visualization capability which facilitates exploratory and traditional analysis of time sequences of three-dimensional data, and (4) a data mining capability closely coupled with the visualization methods.

Title CHRONOS – Integration temporaler Aspekte in ein geographisches Informationssystem
URL http://www.faw.uni-ulm.de/deutsch/projekte/chronos.html
People Wolf-Fritz Riekert
Institution(s) FAW Ulm
Description The project deals with the development of a spatio-temporal data model. Main objective was the design of a spatio-temporal query language for GIS applications. (in German)

Title The Atlas of Historical County Boundaries
URL http://www.newberry.org/nl/scholl/L3ratlas.html
People John H. Long
Institution(s) Newberry Library, Chicago
Description The Atlas of Historical County Boundaries is a project of the William M. Scholl Center for Family and Community History at the Newberry Library in Chicago. It is a reference work designed to provide information about the creation and boundary changes of every county in the United States, from the earliest county creation in the 1600s to 1990.

Title Historisch-Statistische-Grundkarte von Deutschland
URL http://www.htw-dresden.de/vk/4Forschg/Projekte/thudich/start.htm
People Uwe Jäschke
Institution(s) Hochschule für Technik und Wirtschaft Dresden, Fachhbereich Vermessungswesen
Description The project members work on the partial digitization of a historic statistical base map of Germany containing the administrative boundaries at a scale of 1:100,000. The map series called "Thudichum-Karte" was originally edited between 1890 and 1920 and covers almost all territories of the former German Empire. (in German)

Title Project Gigalopolis
URL http://www.ncgia.ucsb.edu/projects/gig/project_gig.htm
People Keith C. Clarke
Institution(s) UCSB, Department of Geographie
Description Gigalopolis is the growing urban structure containing billions of people worldwide. Urban settlements and their connectivity will be the dominant driver of global change during the twenty-first century. Intensely impacting land, atmospheric, and hydrologic resources, urban dynamics has now surpassed the regional scale of megalopolis and must now be considered as a continental and global scale phenomenon. Project Gigalopolis extends and refines the Clarke urban growth model enabling predictions at regional, continental and eventually global scales.

Title COROCHRONOS – A Research Network for Spatiotemporal Database Systems
URL http://www.dbnet.ece.ntua.gr/~choros/
People
Institution(s) European Union
Description The main objective of CHOROCHRONOS is to allow European researchers working on *spatial* and *temporal databases* to achieve a higher understanding of each other's work, integrate their results and methodologies, and advance the state of the art in this area through an intensive three-year research program. This will culminate in the design and partial implementation of an architecture for Spatiotemporal Database Systems (STDBMS).

9.2.2 Conferences

Title Mapping Europe's Historic Boundaries and Borders: An Exploratory Workshop
URL http://www.geog.port.ac.uk/hist-bound/
People Humphrey Southall, Michael Goerke, Gunnar Thorvaldsen
Institution(s) University of Portsmouth, Department of Geography
Description Historical information about states or cities can usually be interpreted using modern maps. However, mapping detailed historical information onto recent base maps causes large errors: over time, administrative boundaries change and placenames evolve. These need detailed research, which can be disseminated as an atlas or, more cheaply and conveniently, as a historical GIS. Both boundary research and GIS construction are necessarily time-consuming and expensive, and can seldom be justified by one researcher's needs, even if they have the skills required. However, such GIS's have many uses.

Title Mapping Europe's Historic Boundaries and Borders
URL http://www.geographie.uni-mannheim.de/hbound/default.htm
People Frank Swiaczny, Humphrey Southall
Institution(s) European Science Foundation, Standing Committee on the Humanities
Description Program, participants and proceedings of the conference held in June 2000

Title Virtual Environments for the Geosciences, Session 3: Visualization of spatio-temporal simulation
URL http://www.itc.nl/enggeo/esf/abstracts/session3.html
Institution(s) European Science Foundation
Description Abstracts of the papers presented.

9.2.3 Publications

Title Development of a Prototype Temporal GIS
URL http://www.gis.bcit.bc.ca/jcandy/thesis/
People Jonathan Thomas Candy
Institution(s) Simon Fraser University, Burnaby B.C., Canada
Description The initial design of the prototype presented in this thesis was driven by a requirement that the standard GIS operates as usual, but has additional TGIS functions. A consequence of this is that all the extra data needed for TGIS feature retrieval and analysis must be

stored within the relational database tables. The design uses a vector update model. The model must support multiple feature versions and multiple attribute versions with associated database time and event time stamps. As feature versions are superseded they are removed from the current data layer and stored in an 'archive' layer. Sufficient information 'travels' with the archived features to make historical data retrieval and query possible. The model is improved by adding feature attributes for historical topology storage. Database tables store TGIS definitions, temporal and non-temporal attributes and event pointers. The prototype TGIS was successfully developed and is based upon Arc/Info GIS software. Macro programs were written to provide menu driven temporal data management, data editing and data display and query functions.

Title Time, points and space - towards a better analysis of wildlife data in GIS
URL http://www.geo.unizh.ch/~imfeld/diss/
People Stephan Imfeld
Institution(s) University of Zurich, Switzerland
Description This thesis is trying to provide a new perspective on how to analyse moving point objects within GIS. A conceptual shift is performed from a space centered view to a way of analysing spatial and temporal aspects in an equally balanced way.

Title Ein objektorientiertes Datenmodell für ein zeitintegratives GIS
URL http://members.tripod.de/Hellwich/Titel.html
People Frank Hellwich
Institution(s) University of Mannheim, Department of Geography
Description This thesis deals with the development of an object-oriented data model for temporal GIS. Main objective is the provision of a "structural skeleton" which may serve as a basis for the development of an OO-temporal GIS. (in German)

Title Temporal Database Bibliography Update
URL http://www.isse.gmu.edu/~csis/tdb/bib97/bib97.html
People Yu Wu, Sushil Jajodia, X. Sean Wang
Institution(s) George Mason University, Center for Secure Information Systems
Description This is the seventh bibliography concerning temporal databases compiled by the authors in 1997. There have been six previously printed publications (detailed references on the website).

Title Of todes and worms: An Experiment in bringing Time to ArcInfo
URL http://www.arcdataonline.com/library/userconf/europroc96/PAPERS/PN35/PN35F.HTM
People Peter Halls & Paul Miller

Institution(s) University of York / University of Newcastle
Description Research use of GIS in a variety of subject areas from Archaeology to Ecology has found the current "time slice" approach to temporal GIS inappropriate to their data, methodology and theoretical tenets. An alternative technique defining a temporal "topology" is now under investigation, using ArcInfo as the development platform. A research project at York is investigating the applicability of this approach in relation to environmental ecology data.

Title The Consensus Glossary of Temporal Database Concepts
URL http://www.cs.auc.dk/~csj/Glossary
Institution(s) Aalborg University, Department of Computer Science
Description The glossary contains definitions and explanations of a wide range of concepts specific to and widely used within temporal databases.

9.2.4 Miscellaneous

Title Temporal GIS - a short introduction
URL http://www.nlh.no/institutt/isf/ansatte/isfss/tgis.htm
People Sverre Stikbakke
Institution(s) Agricultural University of Norway, Department of Forest Sciences
Description This document was made as an addition to a student presentation in the GeoDB course at Department of Mapping Sciences at the Agricultural University of Norway.

Title history-digitisation
URL http://www.mailbase.ac.uk/lists/history-digitisation/
People Paul Ell
Description A mailing list concerned with the digitisation, of historical material whether as an image or an OCRed document. This list aims to bring together experts in the field of digitising historical material and academics who have projects in mind.

Title Förderverein Historische Grundkarte
URL http://www.gfi-gis.com/grundkarte/default.htm
Institution(s) Förderverein Historische Grundkarte e.V.
Description Homepage of a German non-profit society that wants to develop a digital historic base map for Central Europe which documents the alterations of administrative boundaries throughout the 19th and 20th century. (in German)

Title Current Trends and Future Directions in the Development of Temporal GIS
URL http://mscgis.geog.le.ac.uk/97-gjp5/st_outline.html
People Greg Phillips
Institution(s) Leicester University, Department of Geography
Description The study attempts to collate as much relevant background material on the current development of Temporal Geographic Information Systems (TGIS) as possible, and examine the various methods for modelling time in digital information systems.

Title Spatio-Temporal GIS
URL http://homepages.tesco.net/~philip.sargent/spatiotemporalseminar/
People Philip Sargent
Institution(s) Space Applications Institute
Description Seminar prepared in response to discussion after a presentation on OO GIS (1998) when it became clear that the new facilities that the audience wanted from a new-style GIS were largely temporal; rather than the more commonly known advantages of OO GIS.

9.3
CD-ROM with examples and code from the book

For accessing the material on the CD-ROM please open the *index.htm* file with a WWW Browser. Please be sure to check the website at http://www.time-gis.de for additional information and updates.

Index